Production and utilization of products from commercial seaweeds

288

Edited by
Dennis J. McHugh
Department of Chemistry, University College
University of New South Wales
Australian Defence Force Academy
Campbell, ACT 2600, Australia

FOOD
AND
AGRICULTURE
ORGANIZATION
OF THE
UNITED NATIONS
Rome, 1987

The designations employed and the presentation of material in this publication do not imply the expression of any opinion whatsoever on the part of the Food and Agriculture Organization of the United Nations concerning the legal status of any country, territory, city or area or of its authorities, or concerning the delimitation of its frontiers or boundaries.

M-47
ISBN 92-5-102612-2

All rights reserved. No part of this publication may be reproduced, stored in a retrieval system, or transmitted in any form or by any means, electronic, mechanical, photocopying or otherwise, without the prior permission of the copyright owner. Applications for such permission, with a statement of the purpose and extent of the reproduction, should be addressed to the Director, Publications Division, Food and Agriculture Organization of the United Nations, Via delle Terme di Caracalla, 00100 Rome, Italy.

© **FAO 1987**

PREPARATION OF THIS DOCUMENT

This document was prepared and published under the FAO Technical Cooperation Programme as part of the background material for a Workshop on the Production and Utilization of Commercial Seaweeds which was held in Qingdao, People's Republic of China from 25 May to 15 June 1987. The Workshop was funded jointly by FAO, the Academia Sinica and the UNDP.

It was decided to publish it as an FAO Fisheries Technical Paper in order to fill a gap in the available literature on commercial extraction processes of the most important phyco-colloids. The publication is also a companion to FAO Fisheries Technical Paper (281) Case studies of seven commercial seaweed resources, edited by M.S. Doty, J.F. Caddy and B. Santelices.

Distribution:

FAO Fisheries Department
FAO Regional Fisheries
  Officers
Participants in the Workshop
Limited HP Selector

For bibliographic purposes this document should be cited as follows:

McHugh, D.J. (ed.), Production and
1987   utilization of products from commercial seaweeds. FAO Fish.Tech. Pap., (288):189 p.

ABSTRACT

The publication describes the production, properties and main applications of the three major phyco-colloids extracted from seaweed: agar, alginate and carrageenan. There is also a supplementary chapter on the preparation and marketing of seaweeds as food. Although this is based mainly as Japanese experience it is included in order to encourage increased consumption of seaweeds as human food.

The authors of the three phyco-colloid chapters have considered raw material sources, post-harvest treatment, methods of extraction, chemical composition of the colloids, evaluation of quality, and uses, with some details of economics and marketing.

CONTENTS

CHAPTER 1

PRODUCTION, PROPERTIES AND USES OF AGAR 1
by Rafael Armisen and Fernando Galatas

CHAPTER 2

PRODUCTION, PROPERTIES AND USES OF ALGINATES 58
by Dennis J. McHugh

CHAPTER 3

PRODUCTION, PROPERTIES AND USES OF CARRAGEENAN 116
by Norman Stanley

CHAPTER 4

PREPARATION AND MARKETING OF SEAWEEDS AS FOODS 147
by Kazutosi Nisizawa

CHAPTER 1

PRODUCTION, PROPERTIES AND USES OF AGAR

by

Rafael Armisen and Fernando Galatas
Hispanagar, S.A., Poligono Industrial de Villalonquejar
Calle López Bravo "A", 09080 Burgos, Spain

INTRODUCTION

According to the US Pharmacopeia, agar can be defined as a hydrophilic colloid extracted from certain seaweeds of the Rhodophyceae class. It is insoluble in cold water but soluble in boiling water. A 1.5% solution is clear and when it is cooled to $34-43°C$ it forms a firm gel wich does not melt again below $85°C$. It is a mixture of polysaccharides whose basic monomer is galactose. These polysaccharides can be sulphated in very variable degrees but to a lesser degree than in carrageenan. For this reason the ash content is below those of carrageenan, furcelleran (Danish agar) and others. A 5% maximum ash content is acceptable for agar although it is normally maintained between 2.5-4%.

Agar is the phycocolloid of most ancient origin. In Japan, agar is considered to have been discovered by Minoya Tarozaemon in 1658 and a monument is Shimizu-mura commemorates the first time it was manufactured. Originally, and even in the present times, it was made and sold as an extract in solution (hot) or in gel form (cold), to be used promptly in areas near the factories; the product was then known as **tokoroten**. Its industrialization as a dry and stable product started at the beginning of the 18th century and it has since been called **kanten**. The word "agar-agar", however, has a Malayan origin and **agar** is the most commonly accepted term, although in French- and Portuguese-speaking countries it is also called **gelosa**.

A Japanese legend is told about the first preparation of agar:

"A Japanese Emperor and his Royal Party were lost in the mountains during a snow storm and arriving at a small inn, they were ceremoniously treated by the innkeeper who offered them a seaweed jelly dish with their dinner. Maybe the innkeeper prepared too much jelly or the taste was not so palatable but some jelly was thrown away, freezing during the night and crumbling afterwards by thawing and draining, leaving a cracked substance of low density. The innkeeper took the residue and, to his surprise, found that by boiling it up with more water the jelly could be remade".

Agar production by modern techniques of industrial freezing was initiated in California by Matsuoka who registered his patents in 1921 and 1922 in the United States. The present manufacturing method by freezing is the classic one and derives from the American one that was developed in California during the years prior to World War II by H.H. Selby and C.K. Tseng (Selby, 1954; Selby and Wynne, 1973; Tseng, 1946). This work was supported by the American Government which wated the country to be self sufficient in its strategic needs, especially in regard to bacteriological culture media.

Apart from the above American production, practically the only producer of this phycocolloid until World War II was the Japanese industry which has a very traditional industrial structure based on numerous small factories (about 400 factories operated simultaneously). These factories were family operated, producing a non-standardized quality, and had a high employment rate as production was not mechanized. For this reason, and in spite of the later installation of some factories of a medium to small size, only in recent times has Japan operated modern industrial plants.

During the second world war the shortage of available agar acted as an incentive for those countries with coastal resources of <u>Gelidium sesquipedale</u>, which is very similar to the <u>Gelidium pacificum</u> used by the Japanese industry. So in Portugal, Loureiro started the agar industry in Oporto while at the same time J. Mejias and F. Cabrero, in Spain, commenced the studies which led to the establishment of the important Iberian agar industry. Other European countries which did not have agarophyte seaweeds tried to prepare agar substitutes from other seaweed extracts (see Appendix).

SOURCES OF AGAR

Different seaweeds used as the raw material in agar production have given rise to products with differences in their behaviour, although they can all be included in the general definition of agar. For this reason, when agar is mentioned, it is customary to indicate its original raw material as this can affect its applications (Figure 1). Hence we talk about Gelidium agar, Gracilaria agar, Pterocladia agar, etc. To describe the product more accurately, it is usual to mention the origin of the seaweeds, since Gracilaria agar from Chile has different properties from Gracilaria agar from Argentina and Gelidium agar from Spain differs from Gelidium agar from Mexico.

Originally Gelidium agar constituted what we consider genuine agar, assigning the term agaroids to the products extracted from other seaweeds. Although these agaroids do not have the same properties as Gelidium agar, they can be used as substitutes under certain conditions. After World War II, the Japanese industry was forced to use increasing quantities of raw materials other than the traditional Gelidium pacificum or Gelidium amansii due to the growing demand of the international food industry.

An increase in the agar gel strength was obtained through improvements in the industrial process during the fifties, and the differences between the genuine Gelidium agar and the agaroids then available became clearer. The gel strength increased from 400 g/cm^2 (the maximum for natural agar produced by the cottage industry) to 750 g/cm^2 or more for the agar produced by industrial methods. The gel strength data refer to the Nikan-Sui method which replaced the primitive Kobe method used in the past. The Nikan method is more precise and easier to reproduce. A short description of the method is included in the section on "Properties".

The Japanese discovery of the strong alkaline methods for agar extraction (see section on "Manufacturing Processes") meant an increase of the Gracilaria agar gel strength with the subsequent utilization of seaweeds imported from South Africa to increase the raw material available.

Nowadays the world agar industry basically uses the following seaweeds:

(1) Different species of Gelidium harvested mainly in Spain, Portugal, Morocco, Japan, Korea, Mexico, France, USA, People's Republic of China, Chile and South Africa.

(2) Gracilaria of different species harvested in Chile, Argentina, South Africa, Japan, Brazil, Peru, Indonesia, Philippines, People's Republic of China including Taiwan Province, India and Sri Lanka.

(3) Pterocladia capillace from Azores (Portugal) and Pterocladia lucida from New Zealand.

(4) Gelidiella from Egypt, Madagascar, India, etc.

Figure 2 shows Gracilaria and Gelidium

Other seaweeds are utilized as well, such as Ahnpheltia plicata from North Japan and the Sakhalin Islands as well as Acanthopheltis japonica, Ceramiun hypnaeordes and Ceranium boydenii (Levring, Hoppe and Schmid, 1969).

CLASS	SUBCLASS	ORDER	FAMILY	GENUS	SPECIE
RHODOPHYCEAE	FLORIDEAE	GELIDIALES	GELIDIACEAE	GELIDIUM	G.CORNEUM G.CARTILAGINEUM G.PACIFICUM G.PRISTOIDES
				GELIDIELLA	G.ACEROSA
				PTEROCLADIA	P.CAPILLACEA P.LUCIDA
		GIGARTINALES	GRACILARIACEAE	GRACILARIA	G.CORNEUM G.CRASSA G.LICHENOIDES G.CANALICULATA G.LAMANEIFORMIS G.CONFERVOIDES G.VERRUCOSA G.FOLIIFERA G.SJOESTEDII G.GIGAS G.CORTICATA G.MULTIPARTITA
			PHYLLOPHORACEAE	AHNFELTIA	A.PLICATA
		CERAMIALES	CERAMIACEAE	CERAMIUM	C.HYPNEOIDES C.BOYDENII

Figure 1   Rhodophyceae

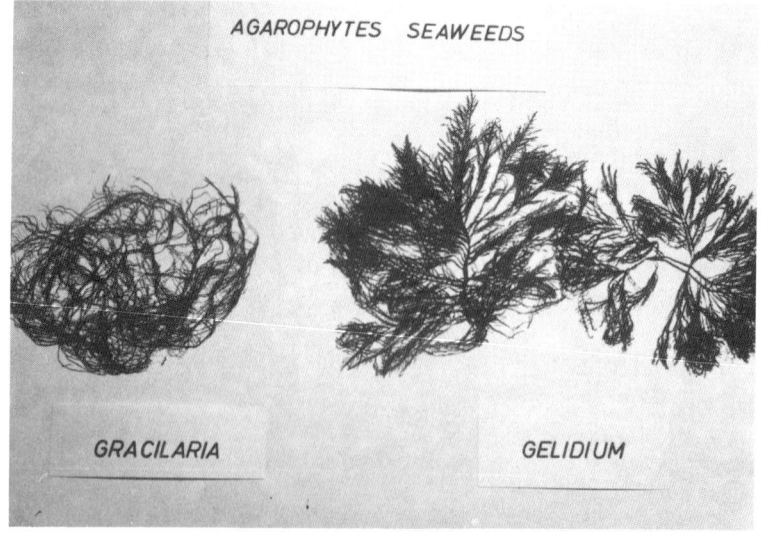

Figure 2   Agarophytes.   Gracilaria and Gelidium

GEOGRAPHICAL DISTRIBUTION

The geographical distribution of agarophytes is very wide and is shown in Figure 3. Main areas are located indicating the most important classes and species. The size of the coloured areas relate to the extent of the gathering area, not the quantity of seaweeds gathered.

There are areas in which different kinds of agarophytes are gathered. This is the case in Chile, a country of exceptional resources of algae. In 1984, 6 126 tons of dried Gracilaria were gathered from its very long sea coast and exported to Japan, as well as another 5 500 tons that were used by the local industry. Simultaneously, in rocky areas, sandwiched between sandy areas where Gracilaria grows, 304 tons of dried Gelidium were gathered and exported to Japan. In countries such as India and Sri Lanka, Gracilaria and Gelidiella grow together in areas relatively close to each other. Generally Gelidium resources are being exploited quite heavily and so are those of good quality Gracilaria. At present the utilization of Gelidiella is being developed.

It is difficult to evaluate the present collection of agarophytes all over the world but since Japan has been, for a long time, the sole importing country of these seaweeds (basically needed to maintain production levels of the agar industry), Japanese statistics (Figure 4a) are very valuable in giving a true view of the situation. Note that in the Japanese statistics, Gelidium seaweeds are separated from other seaweeds. In 1984, Japan imported 678 tons of Gelidium seaweeds and 9 462 tons of "other agarophytes", mainly Gracilaria and Gelidiella. However it seems that Gelidiella is included with Gelidium in some cases, probably because Gelidiella seaweeds have been called Gelidium rigidum by some phycologists in spite of the fact that they are generally considered to be of a different class. As far as agar manufacturers are concerned, they are not Gelidium since the product obtained from then is completely different from the real Gelidium agar. The Gelidium lots assigned to the Philippines (3 tonnes) and to Indonesia (62 tonnes) are probably Gelidiella. Also Gelidium from Brazil is most probably Pterocladia wich can be confused with Gelidium (no Gelidium is harvested in Brazil while some quantities of Pterocladia are).

INDUSTRIAL HARVESTING TECHNIQUES

Industrial harvesting techniques for agarophytes vary, depending on circumstances, but they can be classified as follows:

(1) gathering of seaweeds washed to the shore;

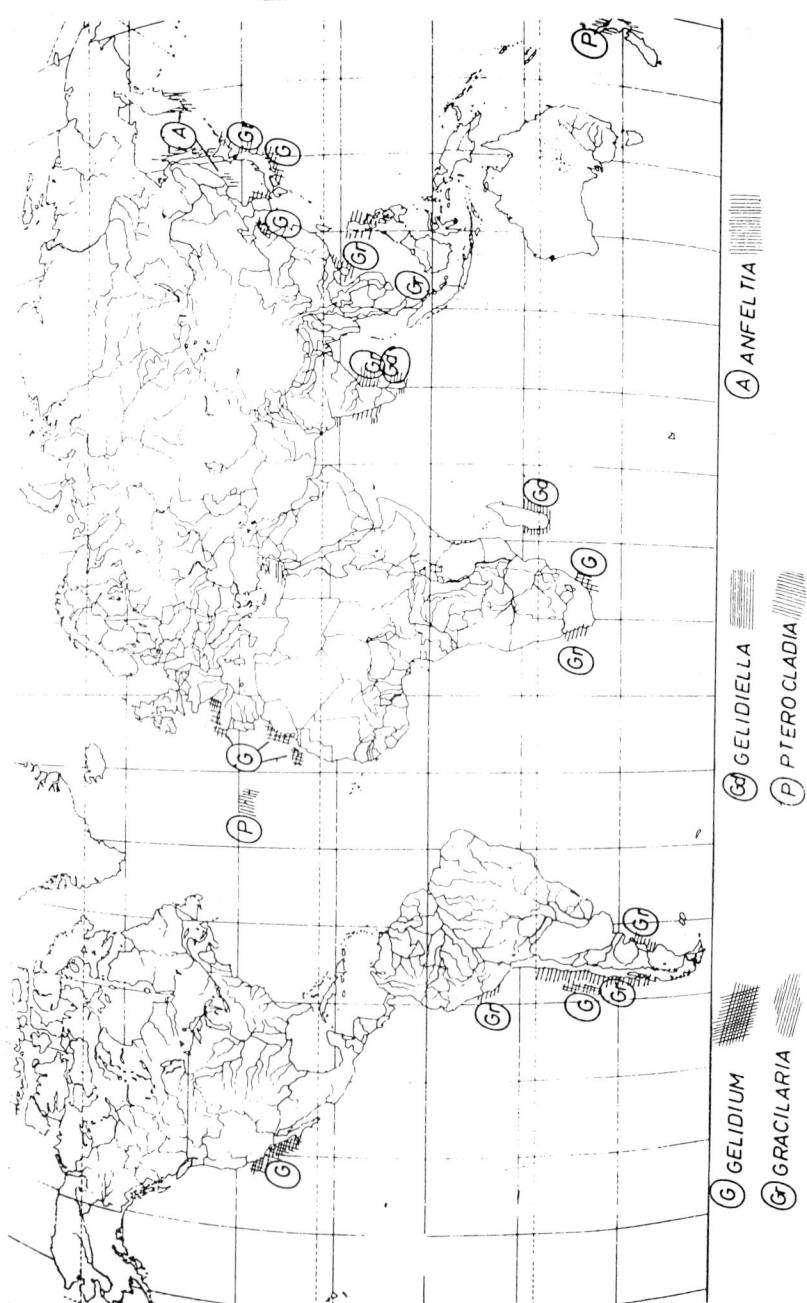

Figure 3 Agarophyte seaweeds distribution map

COUNTRY	QUANTITY

Gelidium Seaweeds:

* North Korea	....	112 MT
* Taiwan	....	4 "
* Philippines	....	3 "
* Indonesia	....	62 "
* Chile	....	303 "
* Brazil	....	20 "
* Madagascar	....	74 "
* South Africa	....	100 "
	TOTAL	678 "

Other Seaweeds:

* North Korea	....	47 MT
* South Korea	....	48 "
* Taiwan	....	77 "
* Vietnam	....	15 "
* Thailand	....	3 "
* Philippines	....	1,470 "
* Indonesia	....	69 "
* Sri Lanka	....	45 "
* Chile	....	6,128 "
* Brazil	....	607 "
* Argentina	....	58 "
* South Africa	....	895 "
	TOTAL........	9,462 "

Figure 4a   Agarophyte seaweeds imported by Japan 1984

COUNTRY	YEAR	GELIDIUM	PTEROCHLADIA	OTHER SEAWEEDS	TOTAL
Japan	1984	568 MT	–	1,872 MT	2,440 MT
Spain	1984	890	–	–	890
Chile	1984	–	–	820	820
South Korea	1984	600	–	–	600
Morocco	1984	550	–	–	550
Portugal	1984	260	60	–	320
Taiwan	1984	25	–	250	275
Argentina	1983	–	–	197	197
Indonesia	1984	–	–	150	150
People's Republic of China	1984	50	–	90	140
Mexico	1984	80	–	–	80
United States ..	1984	70	–	–	70
France	1984	65	–	–	65
Brazil	1983	–	–	60	60
New Zealand	1983	–	26	–	26
TOTALS ..		3,158	86	3,439	6,683

"Other Seaweeds" are practically all Gracilaria
Agar quantities include natural agar, industrial agar and bacteriological agar.
In countries such as India and Vietnam there exists a small agar production but their data are not available.

Figure 4b Agar production in different countries indicating the seaweeds used

(2) gathering seaweeds by cutting or rooting them out from their beds;

(3) cultivation.

Gathering of seaweeds washed to the shore. In some countries these seaweeds called "argazos", "arribazon" or "beach wash". These are dead seaweeds that, after completing their biological cycle, are separated by seasonal storms. They are gathered by hand or by mechanical means from the coast or by compressed air ejectors from boats that gather the seaweeds settled in cavities at depths of about 25 metres ("wells"). To avoid fermentation, the seaweed should be gathered shortly after it has separated from its holdfast.

Gathering seaweeds by cutting or rooting them out from their beds. This work is done with rakes or grabs handled from boats or by scuba divers who operate from boats using compressed air bottles or, more frequently, a compressor on the boat connected to the diver by a hose ("hookah"). Gelidium usually occurs on rocky beds, Gracilaria on sandy ones. In general it is feasible to operate with divers in depths between 3 and 20 metres. In Japan, for many years, the seaweeds have been gathered by diving girls or "amas" who operte from floats and dive using only their lungs. These techniques of cutting or rooting out are used exclusively in some countries and are similar to the ones used for carrageenanophytes such as Chondrus crispus and other Chondrus species (Irish Moss) or alginophytes such as Macrocystis or Laminaria, adapting the equipment in each case to the morphological characteristics of each seaweed class.

Cultivation. Nowadays the need for greater quantities of agarophytes has brought about the introduction of cultivation of Gracilaria crops, along the lines used for carrageenophytes. However this cultivation has had only limited success and there are some aspects to be solved before it can be generally adopted. At the present time, cultivation for industrial purposes is undertaken in the People's Republic of China, its Taiwan Province and it is now being initiated in Chile (Ren, Wang and Chen, 1984; Ren and Chen, 1986; Cheuh and Chen, 1982; Yang, 1982; Pizarro and Barrales, 1986; Santelices and Ugarte, 1986).

POST-HARVEST TREATMENT

The preservation of seaweeds, between the time of harvesting and their actual use by the agar manufacturer, is very important. To build a seaweed processing factory, which consumes seaweeds at the rate they are harvested, is not practical. Large scale agar manufacture makes it necessary to have available quantities of agarophytes stabilized in such a way that they can be carried long distances, at the least possible cost, and stored for a long time before processing.

The first step is preservation through dehydration, to avoid fermentation that first destroys the agar and then the seaweed. The second step is pressing the weed with a hydraulic press in bales of about 60 kg, to reduce the volume and return transportation and storage costs. Dehydration must be sufficient to guarantee the seaweed's preservation, otherwise an anaerobic fermentation will occur inside the bales causing high temperatures and even carbonization of the seaweeds during storage in warehouses. In general, the moisture content is best reduced to about 20% by natural or artificial drying. Obviously it is necessary to avoid wetting during transportation and/or storage.

In the case of Gracilaria the problem is more difficult to solve. The enzymatic hydrolysis of its agar occurs spontaneously even at relatively low moisture contents, but at variable rates depending on the Gracilaria species and its origin. Gracilaria harvested in India, Sri Lanka, Venezuela, Brazil, and generally in warm waters, has an agar (agarose) less resistant to enzymatic hydrolysis than the Chilean Gracilaria which is the most stable. Nevertheless, the stability of agar contained in Gracilaria is less than that of Gelidium; Gelidium agar can be preserved in seaweeds indefinitely provided they have been well treated.

Hydrolysis of agar contained in Gracilaria can be due to endogenous enzymes or to the growth of Bacillus cereus. Pterocladia capillacea from the Azores behaves like Gelidium but the extent of hydrolysis of seaweeds such as Gelidiella, Ahnpheltia, and others has not been described.

SEAWEED PRICES

The seaweeds imported by Japan in 1984 are shown in Figure 4a. Under the heading of "Gelidium", 678 tonnes were imported at a value of Yen 253 741 000 CIF (Yen 374 250 per tonne); the exchange rate at that time was US$ 1 = Yen 246.17 so an average CIF price was US$ 1 520 per tonne. Freight must be considered in this price, especially the high cost of the freight between countries far from Japan such as Chile (303 tonnes), Brazil (20 tonnes), Madagascar (74 tonnes) or South Africa (100 tonnes). These countries account for 73% of the seaweed imported as Gelidium. The seaweeds imported by Japan are high quality and exceptionally clean, otherwise it would not be possible to afford the freight costs. This applies to Gelidium, Gracilaria or any other agarophytes. These specifications, normally not so exact when the manufacturer is located near the harvesting place, greatly increase the harvesting and preparation costs.

The economic data for "Other Seaweeds" is more difficult to interpret because although these seaweeds are referred to as Gracilaria, they may include other agarophytes like Gelidiella, Pterocladia, Ahnpheltia, etc., with different agar contents and therefore different properties. Under the heading of "Other Seaweeds", 9 462 tonnes were imported (compared with Gelidium at 678 tonnes) with a value of Yen 2 882 525 000 (Yen 304 642 per tonne) or US$ 1 237 per tonne (CIF). The freight costs in this price must also be considered since more than 80% of the "Other Seaweeds" are imported from countries such as Chile (6 128 tonnes), Brazil (607 tonnes), Argentina (58 tonnes), and South Africa (895 tonnes).

Gracilaria is the major component of "Other Seaweeds". However two types of Gracilaria are imported and they are not distinguished from each other in the import statistics. One type is clean, dried seaweed. The other type is Gracilaria which has been subjected to a strong alkaline treatment in the exporting country; this causes alkaline hydrolysis of sulphate groups, increasing the gel strength of the agar which is eventually extracted, although the yield is reduced. This treatment is expensive and for this reason the product obtained ("Colagar") brings a higher price than the untreated dried seaweed. All of this must be borne in mind when considering the average price of the "Other Seaweeds" which is calculted from the Japanese import statistics.

EVALUATION OF AGAROPHYTES

The existing literature on the evaluation of seaweeds as industrial sources of agar is confusing because in general the contributions have come from well intentioned scientists who often are unfamiliar with specification requirements, the different grades of commercial agar and the analytical methods used.

Firstly, the evaluations have been made from seaweeds which are perfectly dry and clean, like herbal samples, and therefore their data does not have any similarity to that obtained by the manufacturers who process hundreds or thousands of tonnes of commercial seaweeds arriving in very different stages of preservation, often mixed with significant quantities of impurities such as stones, shells, sand, other seaweeds, epiphytes, as well as other products added during gathering, drying, and packing (such as land weeds, leaves, wood, plastics, etc.). The quantities of salts that remain in the seaweeds after drying is another variable.

Secondly, the evaluation is frequently made without taking into account the characteristics of the agar obtained, or by comparing it only in some parameters (for instance with a bacteriological agar sample). After making some tests without knowing the real specifications for the products, and in many cases without knowledge of the usual analytical methods, it is declared to be similar. All of this is equivalent to determining the density of a rock and, seeing that it is 3.52, deducing that the rock is diamond.

Thirdly, an agarophyte evaluation is a much longer and complicated process than the one usually carried out and published in scientific articles. In our opinion it must be initiated with a series of tests or experiments on a laboratory scale. In principle we would make several extractions, combining some preliminary treatments with different extraction conditions. The previous experience of the person who is going to chose the experiments is very important if good results are to be obtained. Under these conditions it is usual to operate with glass equipment and with quantities of about 50 g of seaweed for each test. For normal work it is necessary to have between 400 and 500 g of dried seaweed.

Next, provided promising results have been obtained and a simplified quality control test has been performed, a pilot plant run should be the next step. An evaluation performed in a laboratory can be sufficient for a scientific publication but in industry, before working in a factory, we operate a pilot plant trial with quantities between 750 g and 1 kg of dried seaweeds in conditions as similar as possible to those of the industrial process. To carry out this trial it is convenient to have about 5 kg of dried seaweed available. Using the agar obtained in the pilot plant, complete analyses are made, as well as an evaluation of the product in practical applications. Knowledge of industrial manufacturing processes for agar is needed for this evaluation to be useful, as well as experience of actual specifications required by the different markets and by the practical applications of the product.

It is essential that agarophytes be correctly evaluated before starting operations in those countries that at present are studying their algal resources. For this, as soon as the quantities of agarophytes from a part of the coast have been estimated, even approximately, the quantity and quality of the agar in the seaweed should be evaluated in terms of its practical use. For this purpose it is important to have the cooperation of experienced agar manufacturers.

A very important point to be considered is the way representative samples are taken from large areas of agarophytes. Sampling is not as easy as it may seem. To have representative samples it is necessary to follow the classical sampling procedures and take some additional special precautions. The sample must be immediately packed in strong, waterproof, well fastened bags; too often samples are received in broken packages containing extremely dried seaweeds and in many cases with significant quantities of sand outside the bag and spread through the package.

As soon as the sample is received in the control laboratory, the impermeability of the plastic bag is verified and registered in the protocol. Next, on aliquots taken in such a way that their homogeneous composition is guaranteed, the following determinations must be made.

1. __Moisture.__ Use a drying oven at 65°C.

2. __Pure seaweed determination.__

   Seaweeds must be soaked in fresh water for at least two hours, with stirring, to eliminate soil and sand which are decanted, filtered, dried and weighed separately. Once the seaweeds are fully swelled the agarophytes must be manually separated from all the other materials such as rocks, shells, calcareous inclusions, other seaweeds, epiphytes, various vegetable remains, wood, plastic, etc. All these materials must be dried and weighed. Then the agarophytes are washed with water until clear (some samples, particularly __Gracilaria__, may contain clay). When cleaned they must be dried (in an oven at 65°C) and weighed; the percentage of the sample which is "pure seaweed" is calculated.

3. __Extraction - of an aliquot part of the sample__

   It is impossible to assign a general extraction method valid for any agarophyte. For many years we have been industrially evaluating a large number of agarophyte batches. They have come from the five continents and include __Gelidium__, __Gelidiella__, __Pterocladia__ and __Gracilaria__ species. We do know that it is impossible to give a valid extraction method for any agarophyte to evaluate its yield, obtaining at the same time a standard quality which allows evaluation to be useful from the point of view of the industry.

   Nevertheless, we shall try to give here certain procedures that are only evaluation methods and should not be confused with industrial processes.

   It is advisable to use 50 g or more in case the agarophytes have a lower percentage of "pure seaweed" than usual. Extraction conditions (pH, temperature, pressure, time, etc.) as well as the seaweed: water ratio must be adapted in each case so as to try to obtain an extract with approximately 1% agar.

   A traditional Japanese method for __Gelidium__ is the following. The seaweed (40 g) is washed three times. It is then placed in a beaker with water (40 ml, or more if necessary, to cover the seaweed which can be flattened). Adjust to pH4 using acetic acid. After 10 minutes the temperature is increased and maintained close to the boiling point for three minutes. Water is added to bring the total volume to 800 ml; with this dilution the pH increases to approximately 6 unless it is adjusted with acetic acid or a dilute solution of caustic soda. The extraction is carried out at a temperature just below boiling point for 3-4 hours, checking the seaweed texture to determine the end of the extraction. The liquid is then filtered through a cloth and the residue is squeezed. As soon as the extract gels, it is subjected to freezing, or syneresis, and afterwards is dried and weighed.

In general Gelidium, Pterocladia or Gelidiella seaweeds can also be evaluated as follows. The seaweed is mixed with a solution of sodium carbonate (0.5%, 30 ml solution for each gram of seaweed) and held at approximately 90°C for 30 minutes, allowing the alkali to diffuse into the seaweed. The seaweed is washed with running water for 10 minutes. It is then extracted with water, 30 ml for each gram of seaweed, adjusting the pH with tartaric acid between 4.8-8.0 depending on the type of seaweed and on the extracting conditions. If extraction is done at boiling point (without pressure) it is usual to work between pH 4.8-6.0 and if it is done under pressure it will depend on the pressure used but at 127°C (1.5 kg.cm^{-2}) it is usual to operate between pH 6-8. The solution is filtered and the product is finished, either by freezing or syneresis, and dried.

In the case of certain Gracilaria species, it is necessary to make what is called a sulfate alkaline hydrolysis, working in stronger alkaline conditions to change the L-galactose 6-sulfate into 3,6-anhydro-L-galactose. For this purpose the diffusion is made with sodium hydroxide solution (at least 0.1M) for at least one hour at a temperature between 80-97°C, but with care not to extract the agar. The extraction which follows is carried out with stirring at practically neutral pH, without pressure (95-100°C), for a very variable time depending on the type of the Gracilaria used, but it can take several hours.

Then analytical control test will be needed to verify that the agar obtained meets the physico-chemical specifications that will be explained later.

CHEMICAL STRUCTURE

Early studies of agar showed that it contained galactose, 3,6-anhydro-galactose (Hands and Peats, 1938; Percival, Somerville and Forbes, 1938) and inorganic sulfate bonded to the carbohydrate (Samec and Isajevic, 1922).

Structural studies have been based on the fractionation of agar by several methods, followed by chemical and enzymatic hydrolysis. The enzymatic hydrolysis studies of W. Yaphe have been of great importance. Subsequently the spectrochemical studies using infrared spectroscopy and nuclear magnetic resonance spectroscopy, particularly ^{13}C n.m.r., have explained many important points in the structure of these intricate polysaccharides.

Infrared spectroscopy is the most accessible method for many laboratories. Figure 8a shows different absorption bands that have been characterized for the agar spectrum. The typical bands of a carrageenan spectrum are also shown (Figure 8b) because many of its important uses are similar to those of agar and the spectra are useful for distinguishing the two. The bands at 1 540 and at 1 640 cm^{-1} are especially noteworthy. They come from the proteins existing in agar and about which only a few comments have been made before. The peak at 890 cm^{-1} has not been identified up to the present time.

N.M.R. is of great importance when studying these structures. However the technique is difficult and it requires ^{13}C n.m.r. equipment which only a few laboratories can afford. For this kind of work it is best to consult W. Yaphe's papers, published from 1977 - for example, Bhattacharjee, Hamer and Yaphe, (1979); Yaphe (1984); Lahaye, Rochas and Yaphe (1986).

Agar is now considered to consist of two fractions, agarose and agaropectin. These were first separated by Araki (1937) and the results were published in Japanese so they were not readily available to some research workers. For example Jones and Peats (1942) assigned a single structure to agar defining it as a long D-galactose chain residue, joined by 1,3-glycosidic links; in the proposed structure, this chain was ended by a residue of L-galactose joined to the chain at C-4 and with C-6 semi-esterified by sulfuric acid. This false structure is still mentioned in some books on natural polymers and even in recently published encyclopedias.

AGAROSE

Interest in agarose was lost until Hjerten, working under Tiselius at the University of Uppsala, began to look for an electrically neutral polysaccharide suitable for electrophoresis and chromatography. He published an improved method of separation based on the use of quaternary ammonium salts (Hjerten, 1962). A technique for agarose preparation using polyethylene glycol was reported by Russell, Mead and Polson (1964) and later this was patented with Polson (1965) named as the inventor. Both methods gave agarose of sufficient purity to allow the study of its structure.

Figure 5 shows the type, and approximate relative quntities, of the residues that can be separated from the total hydrolysis of agarose.

Figure 6 shows agarose to be a neutral, long-chain molecule formed by β-D-galactopyranose residues connected through C-1 and C-3 with 3,6-anhydro-L-galactose residues connected through C-2 and C-4. Both residues are repeated alternately. The links between the monomers have different resistance to chemical and enzymatic hydrolysis. 1,3-α links are more easily hydrolysed by enzymes (Pseudomonas atlantica) and neoagarobiose results. 1,4-β links are more easily hydrolysed by acid catalysts and yield agarobiose units. Nevertheless 1,4-β links make the polysaccharide chain particularly compact and resistant to breakage, as is found in the peptidoglycan of bacteria. The molecular weight assigned to non-degraded agarose is

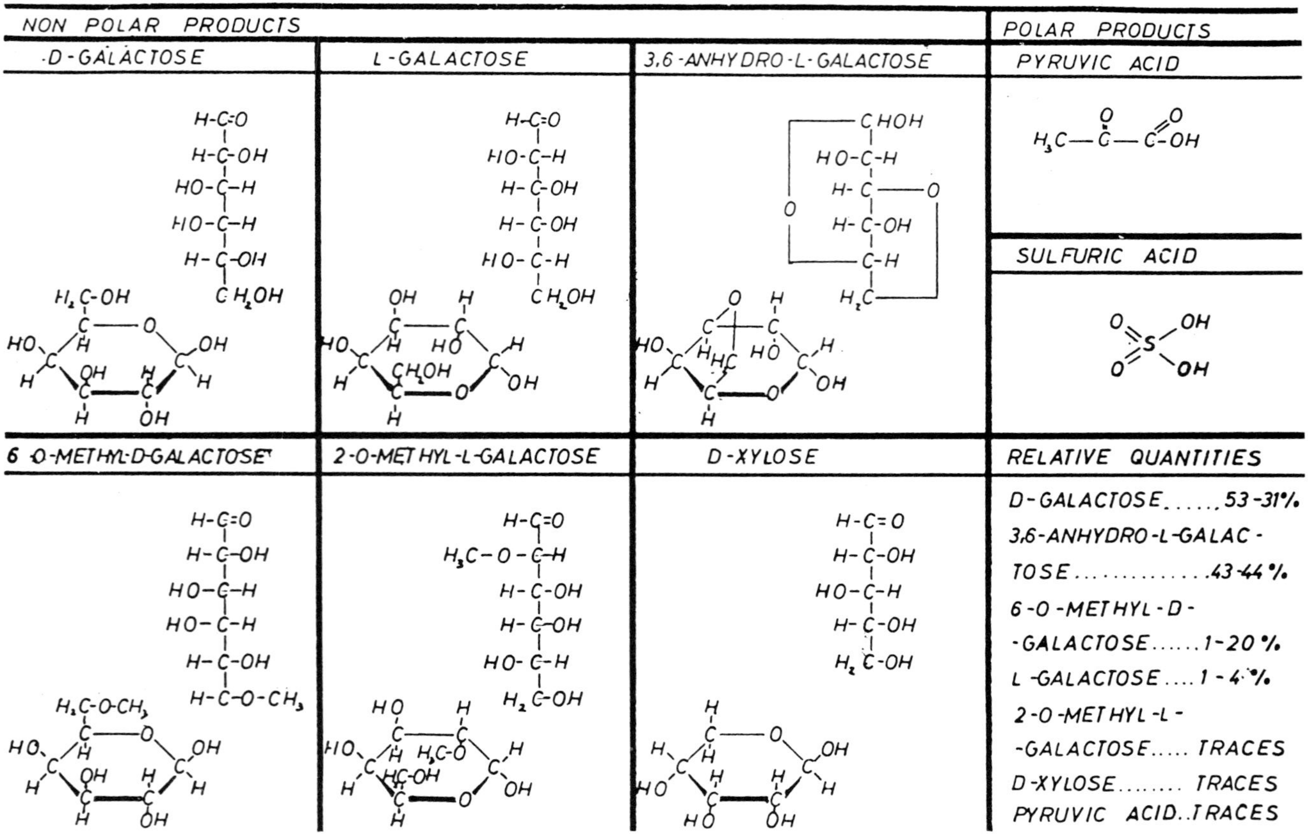

Figure 5 Agarose hydrolysis products

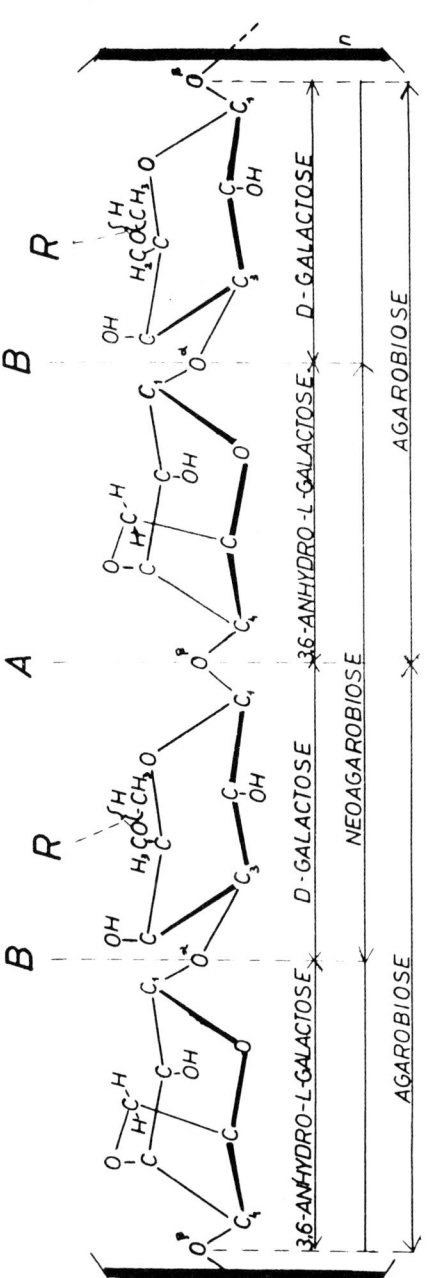

A LINK EASILY HYDROLYSED BY ACIDS.
B LINK EASILY HYDROLYSED BY ENZYMES.
R SUBSITUTION OF -H BY -CH₃ IS CONDITIONED BY THE SEAWEEDS USED.
GEL POINT IS INCREASED BY A HIGHER CONTENT OF -CH₃ GROUPS.
n POLYMERIZATION GRADE. IT CONDITIONS GEL STRENGTH.

Figure 6 Agarose structure

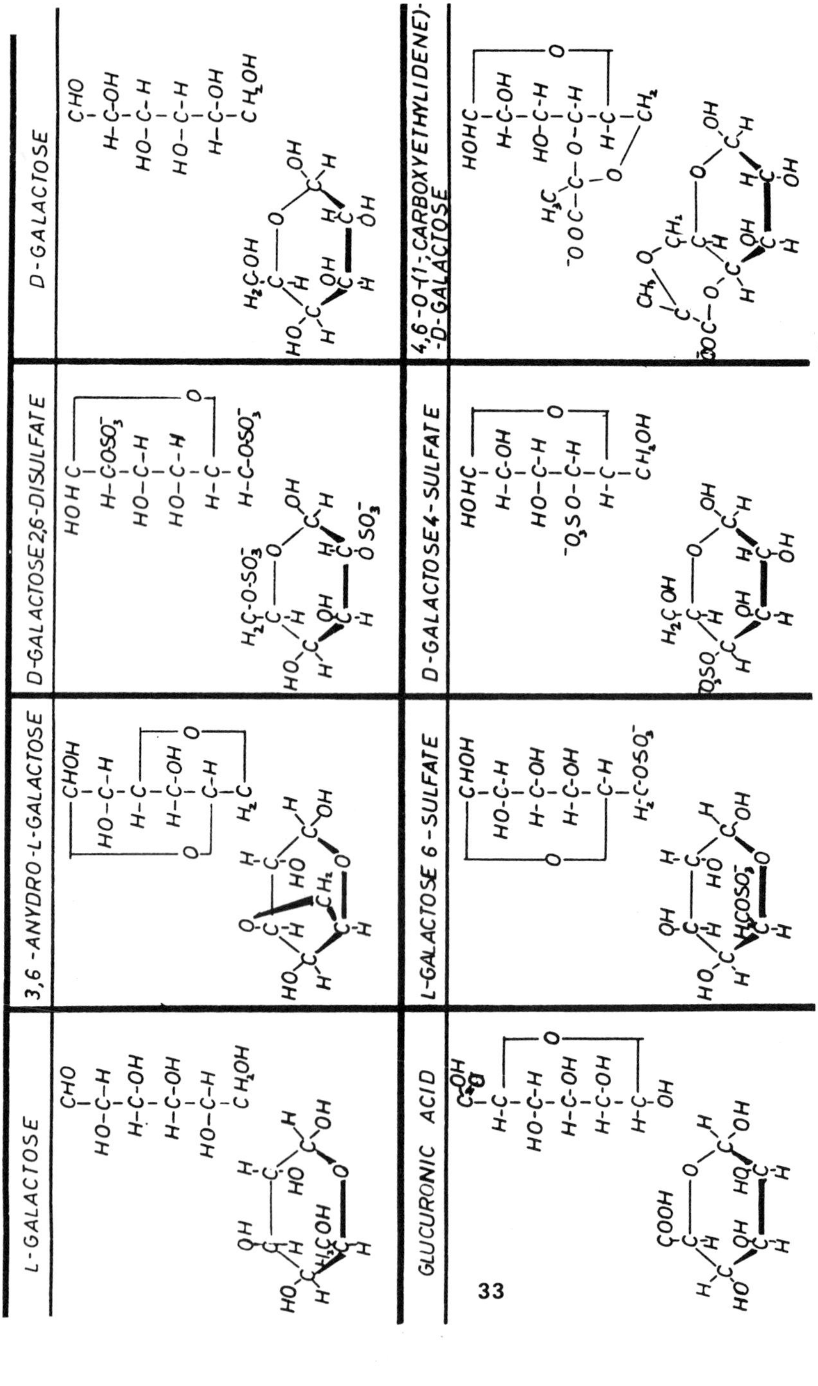

Figure 7 Agaropectins hydrolysis products

WAVE NUMBERS	STRUCTURE CAUSING ABSORPTION
730	Carbon-sulfur links vibration. (Cross, 1964).
750	Carbon-sulfur links vibration. (Torres-Pombo, 1972).
820	Ester-sulfate in C-6 link vibration. (Stancioff and Stanley, 1969).
850	C-O-S in C-4 link vibration. (De Lestang and Lloyd, 1961; Alkahane and Izumi, 1976).
890	Typical Agar peak with unkown meaning.
930	3,6-Anhydro-galactose bridge vibration. Typical Agar peak. (Stanley, (1963)
1060	Ester-sulfate link vibrations. (Cross, 1964). (1)
1070	3,6-Anhydro-galactose bridge vibration. Typical Agar peak. (Stanley, (1963)
1180	Ester-sulfate link vibrations (Cross, 1964). (1)
1250	Ester-sulfate link vibrations, (Alkahane and Izumi, 1976). (1)
1370	Ester-sulfate link vibrations. (Cross, 1964). (1)
1410	Peak with unkown meaning.
1540	CO-NH peptide link vibrations.(Cristiaen, 1983).
1640	Amine function deformations vibrations. (Cristiaen, 1983).
1750	Possibly a methyl group vibration. (2)
2815	O-CH$_3$ link vibrations.
2830	O-CH$_3$ link vibrations.

NOTES:

(1) Peaks at 1060, 1180, 1250 and 1370 are produced by sulfates but the position occupied in the chain by the sulfates is not clearly seen in Agar due its low content of sulfates (< 2%).

(2) Peak at 1750 not attributed up to this moment could be caused by methyl groups as Agar with 6-methyl forms a peak at 1780 cm^{-1}.

Figure 8a Infrared spectrocopy on agar films

			ABSORBANCY RELATIVE TO 1050 CM^{-1}		
WAVE NUMBER		MOLECULAR ASSIGNMENT	KAPPA	IOTA	LAMBDA
800 -	805	3,6-ANHYDRO-GALACTOSE 2-SULFATE. ($-O-SO_3^-$ axial vibration on C-2 of a 3,6-anhydro-galactose ring).	0 -0'2	0'2-0'4	-
810 -	820	GALACTOSE 6-SULFATE	-	-	0'1-0'3
825 -	830	GALACTOSE 2-SULFATE	-	-	0'2-0'4
840 -	850	GALACTOSE 4-SULFATE ($-O-SO_3^-$ axial vibration on C-4 of galactose.)	0'3-0'5	0'2-0'4	-
928 -	933	3,6-ANHYDRO-GALACTOSE (Anhydro-galactose -C-O vibration).	0'3-0'6	0'2-0'4	0' -0'2
1.220 -	1.260	ESTER SULFATE (-S=O vibration).	0'7-1'2	1'2-1'6	1'4-2'0

NOTES.-

1.- A peak at 831 CM^{-1} wide is mentioned in the Bibliography to correspond to a 3-Sulfates mixture.

   $-O-SO_3^-$ equatorial vibration on C-2 of a galactose linked in (1 → 3) ring.

   $-O-SO_3^-$ vibration on C-2 of a galactose linked in (1 → 4) ring.

   $-O-SO_3^-$ vibration on C-6 of a galactose ring.

2.- Carrageenans have wide and strong absorption bands in 1,000-1,100 CM^{-1} region which are typical in all polysacharides.

3.- Maximum absorption is given by 1,065-1,020 CM^{-1} for all carrageenan types (Kappa, Iota, Lambda, etc.)

Figure 8b   Infrared spectrocopy on carrageenan films

approximately 120 000. This weight has been determined by sedimentation measurements and it represents 400 agarbiose (or 800 hexose) units linked together.

This clarifies the information in Figure 6. However it should be noted that, depending on the origin of the raw material, some units of 3,6-anhydro-L-galactose are replaced by L-galactose. Also some D-galactose and L-galactose units can be methylated and it is said they can be in fact 6-0-methyl-D-galactose and 2-0-methyl-Lgalactose. This methylation, arising from the seaweed used in the process, determines the agarose gel point and therefore that of the agar it comes from. D-xylose has been found in very small quantities from hydrolysed agarose but it has not been possible to assign it a position in the structure.

Polar residues such as pyruvic and sulfuric acids are also found in small quantities. They may come from the small amounts of agaropectin lef in the agarose after its preparation but in our opinion sulfate and pyruvate groups remain linked in small quantities to the agarose structure, depending on the seaweed used in agar production. We follow the traditional definition of agaroses as those products obtained as the non-charged fraction after using a classical separation technique such as the precipitation with quaternary ammonium salts by Hjerten. On the other hand, in spite of the copious bibliography on this matter (we have seen 14 different basic methods to prepare agarose), none of the methods permits an agarose preparation free of electronegative charges. Many researchers have used two or three fractionating methods successively, in order to improve the separation and reduce the amount of electronegative groups present. In spite of all these efforts, these groups could not be eliminated. To cancel the electroendosmotic flow, which might be induced by these electronegative groups, it has been necessary to fix electropositive groups or use some other means so as to reduce the migration of cations (and their solvation water molecules) fixed to electonegative groups. Consequently we consider the agarose theoretical structure a chimerical dream to which we get closer each time by using more refined fractionation methods although perhaps, in practice, it may not exist at all in agar and the agarophyte seaweeds.

Nowadays commercial agaroses for use in biochemical separation techniques have to be chemically modified, so that their structure is different from the agarose as it is extracted from the seaweed. Phycologists should be aware that this is so, unless the manufacturer states that the original chemical structure has not been modified.

Agarose is responsible for the gelling power as we know it in agar. This is a gelation in aqueous media with a very small reactivity with cations and proteins and this differentiates agar from carrageenan.

AGAROPECTIN

Agaropectin (or better, the agaropectins) have a low gelling power in water. At the present time, a specific structure has not been assigned to the agaropectins. It is customary to say that they are formed by alternating units of D-galactose and L-galactose, and that they contain all the polar groups existing in agar.

Figure 7 shows the residues obtained by hydrolysis; among them, sulfated and pyruvate residues are evident. It has been verified that L-galactose 6-sulfate and D-galactose 4-sulfate are the major sulfate residues in agar. From small to moderate quantities of 3,6-anhydro-L-galactose have also been detected. These small quantities vary depending on the origin of the seaweed, on the harvesting season, on the treatment applied during the agar manufacturing process and on the treatment used during the agarose separation process.

The presence of 4,6-0-(1-carboxyethylidene)-D-galactose has also been verified, making the position of pyruvic acid in the structure perfectly clear. This unit is relatively important in agaropectin but in agarose it appears in much lower levels, as mentioned previously, probably because agarose has terminal units of 4,6-0-(1-carboxyethylidene)-D-galactose. The quantity of pyruvic acid in agar and agarose varies widely depending on the seaweeds used as raw material; we have verified quantities between 0.2-2.50% in agar and 0.02-1.30% in agarose. In this regard the work of Hirase (1957) is very interesting and explanatory.

These variations, that sometimes can be very important, appear even in seaweeds of the same class harvested a short distance from each other and seem to be permanent and depend on the growing locations. Over a period of several years (more than 10 in some cases) we have studied different Gelidium or Gracilaria harvesting areas in Europe, Asia and America, verifying the persistence of this phenomenon that can be caused by microclimatic differences. In our opinion the differences in cations existing in certain habitats also can be a cause. Naturally the different types and species cause differences that are very important sometimes in the agarose and agaropectin structures.

In Figure 7, D-galactose 2,6-disulfate has been included because we think we have identified it in small quantities in the agaropectins of some seaweeds grown in difficult conditions ("El Niño" phenomenon). These agaropectins had high viscosity, that was also apparent in the agar from which they came, along with a lower gelling power. In cases where this sulfated residue is found, the agaropectin and the agar have undesirable properties. Also shown in Figure 7 are D-galactose and L-galactose which appear to be modular units of agaropectin. Glucuronic acid is present only in traces (like the D-xylose found in agarose).

So while the basic structure of agaropectin consists of alternating D-galactose and L-galactose, D-galactose can be substituted by D-galactose 4-sulfate, by 4,6-0-(1-carboxyethylidene)-D-galactose in certain terminal chain positions or even possibly by D-galactose 2,6-disulfate, while part of L-galactose can be replaced by 3,6-anhydro-L-galactose. These different substitutions of the basic monosaccharide give an enormous number of possible structures.

McCandless used an immunochemical method to detect different carrageenan fractions with great sensitivity (Di Ninno and McCandless, 1978 and 1978a). A similar method might be applied to studying the different kinds of agaropectins in regard to their different seaweed origins, as well as the posible evolution of the structure of agaropectins during the life of the seaweed. To do this it is necessary to take into consideration the different fractions preceding the series of biochemical transformations produced by the algal enzymatic mechanisms which result in certain terminal fractions (one of which may be agarose). The current possibilities through monoclonal antibodies would allow an improvement of the sensitivity and selectivity of the method used by McCandless.

MANUFACTURING PROCESSES

The production of agar, bacteriological agar and agarose are considered in this section.

AGAR

Agar manufacturing processes have developed since the early freezing method was used to concentrate the extracts of agarophyte seaweeds. Whichever process is used, the following criteria should be taken into consideration. Firstly, it is necessary to obtain an extract from agarophyte seaweeds that contains the largest possible amount of the existing agar in the agarophytes. Secondly, the agar obtained should have the best possible characteristics to satisfy the standards expected for this product, especially as far as the gel strength is concerned. To achieve this it is necessary to consider the following basic points for the manufacturing process.

1. The seaweed treatment prior to extraction.

2. The control of molecular weight distribution during the extraction.

3. The removal of undesired products.

4. The need to work with large volumes of dilute extracts.

5. The economics of dehydrating the dilute extracts.

1. SEAWEED TREATMENT PRIOR TO EXTRACTION

The seaweed treatments prior to extraction are very important as they will condition to a high degree the characteristics of the agar obtained. For example Gracilaria agar was once called an agaroid because at that time Gracilaria was not preteated properly resulting in a product softer than that obtained from Gelidium. Now Gracilaria is given a strong alkaline treatment before extraction. This causes hydrolysis of sulfate groups and transforms important quantities of L-galactose 6-sulfate into 3,6-anhydro-L-galactose, thereby significantly increasing the gel strength of the agar obtained. Tagawa and Kojima (1972) say the industry uses 0.25-05M sodium hydroxide solution at 80-90°C for 3-5 hours. Okazaki (1971) gives more detail, showing how the treatment varies depending on the country of origin of the Gracilaria (Okazaki is a useful reference for details of all methods used in the Japanese agar industry). Yang (1982) gives references to the methods used in Taiwan Province. The treatment, also called sulfate alkaline hydrolysis, must be adapted to the class of seaweed used, to obtain as much desulfation as possible while still avoiding the yield losses that this process can cause. These losses can be very important if agar is dissolved in the alkaline solution. The way these treatments are applied is variable and constitutes a part of the manufacturing process that has to be constantly adapted, according to the changing seaweeds, as it becomes a double-edged tool that can substantially reduce the yield if it is wrongly applied.

2. CONTROL OF MOLECULAR WEIGHT DURING EXTRACTION

Agar, as it occurs in seaweed, when extracted is insoluble in cold water and also practically insoluble in hot water. It is therefore necessary to extract it using suitable pH and redox conditions so that some hydrolysis occurs, thereby increasing its solubility. During this fractionation or cracking, it is necessary to avoid the subsequent reduction, by hydrolysis, of the molecular weight of the fragments which have dissolved. As all manufacturing methods are based on agar being soluble in hot water but insoluble in cold, excessive molecular weight reduction of the agar in solution would cause reduction of yields during the process, whenever molecular weights are reached for which cold solutions are possible. On the other hand it is important to avoid molecular units, in the agarophyte residues, that are not soluble either for lack of the necessary solution time or because of an excessive molecular weight that curtails solution under the conditions of extraction.

Figure 10 attempts to clarify a complex process in a simplified way since what we are putting into solution is not only agarose, with a quite uniform chemical structure, but also a mixture of agaropectins carrying electronegative charges, with a minimum solubility temperature that is above the one for agarose. We can see in the figure that all those molecules with molecular weights below PM1 will be easily extracted from the seaweed but will be lost due to their cold water solubility. In contrast, those molecules that remain in the seaweed with molecular weight above PM2 will not be extracted and will remain with the cellulose residues after extraction. The agar manufacturer has to establish working methods that enable the preparation of a molecular weight distribution curve that avoids both losses as much as possible. An ideal result would be that shown by the middle graph of the three shown in Figure 10. It is very difficult to modify the PM1 value but it is possible to increase the PM2 by raising the water temperature in the extraction; this is done by working under pressure whenever the seaweeds permit it. Naturally the differing stabilities of agars to hydrolysis poses limits to such temperature increases.

The industrial objective aims toward narrowing the type of Gaussian curve shown in Figure 10. This reduces losses and increases the molecular weight to the corresponding maximum in the chart which is accompanied by an increase in the agar gel strength. Such considerations will be correct whenever a constant agarose-agaropectin ratio is maintained.

3. REMOVAL OF UNDESIRED PRODUCTS

During the extraction process, a myriad of undesired products will be obtained as well as agar. Such products are soluble salts, seaweed pigments, cellulose, hemicellulose and many extracts coming from impurities and foreign materials contained in the weed, since commercial seaweeds differ greatly from those with which scientists work. Therefore in order to obtain the purest possible extracts in industry, seaweeds are selected and washed carefully and subjected to previous corrective treatments in which generally an alkaline solution eliminates a large quantity of foreign substances, particularly red pigments (phycoestrine), changing the weed to a green colour. This alkaline treatment is with sodium carbonate; it is milder than the alkaline treatment with sodium hydroxide which is used to improve the gel strength of Gracilaria agar.

A careful filtration will purify the extract but this is quite a difficult operation which requires a high temperature (85-100°C) because of the extract's viscosity and high gelling power. Also cellulose and seaweed "floridean starch" residues, and even clay particles, make the filtration very difficult. Pressure filters are commonly used. Filter presses are the most useful ones, although modern factories use filters specially designed for this purpose.

Differences in the raw material greatly influence the operating methods and this makes further generalizations impossible.

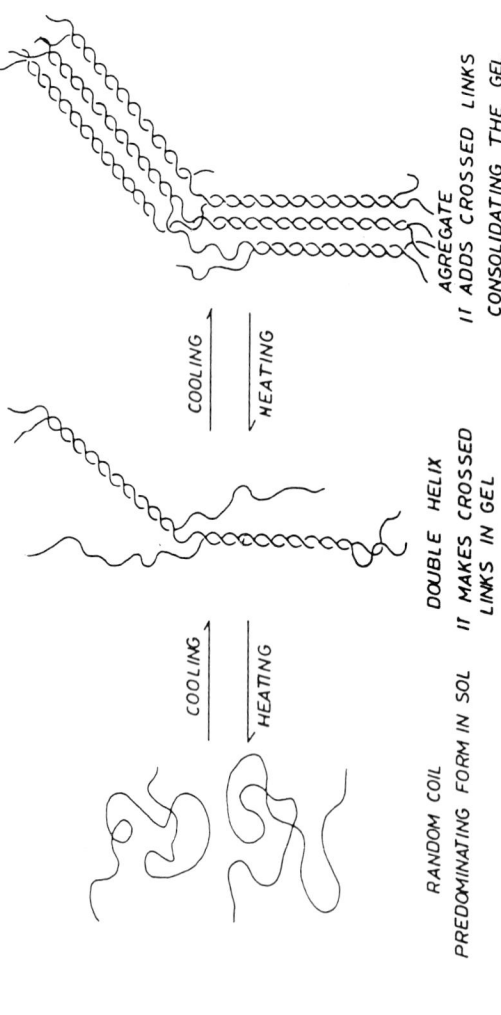

Figure 9 Agarose gelification

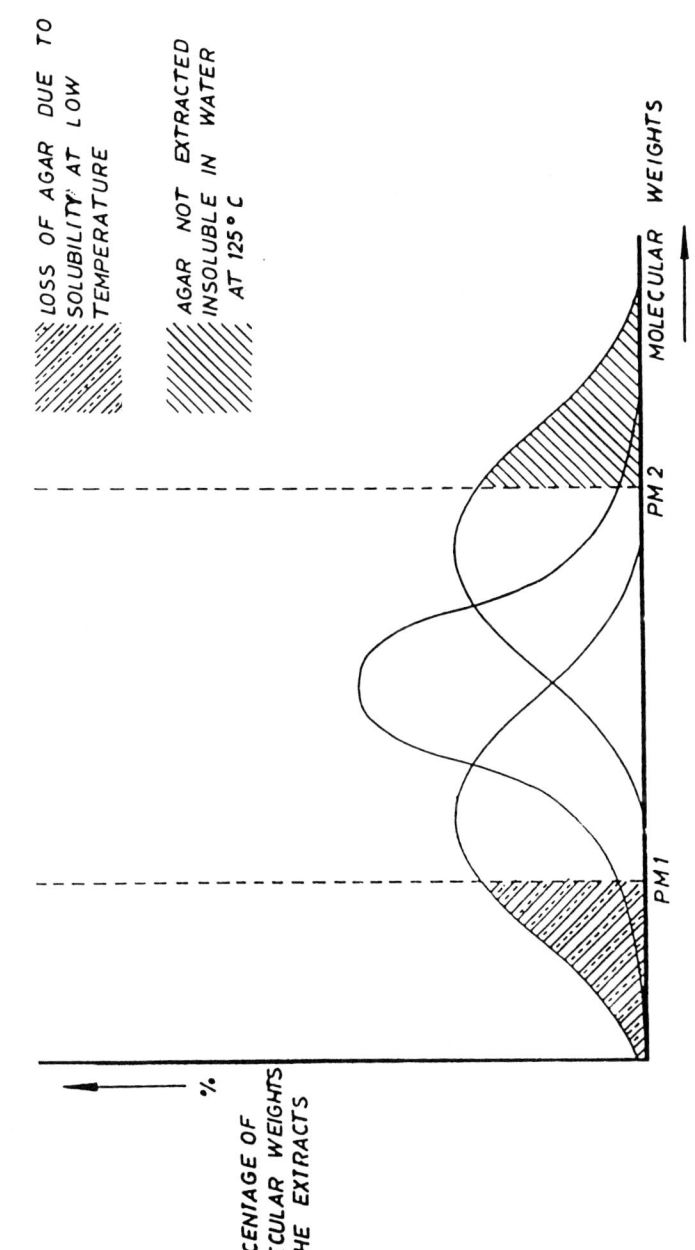

Figure 10 Distribution of molecular weights in agar extracts

4. LARGE VOLUMES OF DILUTE EXTRACTS

Due to the high seaweed cost, high yields of agar are essential. However the extract concentrations range from 0.8% to 1.5% as a maximum; it is difficult to work with a more concentrated extract, for filtration as well as in the rest of the process. So the more agar that is extracted, the more water must be added to keep the concentration in the above range. This means that it is necessary to work with large volumes of extracts.

5. THE ECONOMICS OF DEHYDRATING THE DILUTE EXTRACTS

An important aspect to consider is the economics for dehydrating the large volumes of dilute extracts discussed in (4). This is a characteristic problem for this industry and its solution lies in methods based on the insolubility of agar when the extracts are cooled. Sometimes, because of lack of experience with the industry, projects are encountered in which evaporation or precipitation are recommended as the means of removing the large quantities of water from the extracts. We would first like to show why these methods are not feasible and afterwards discuss the methods actually used by the industry.

A. EVAPORATION

Starting from a 1% extract, 99 litres of water have to be eliminated for each kilogram of agar and since the latent heat enthalpy for water at 100°C is 539 kcal/kg, we need 53 361 kcal/kg (539 x 99 = 53 361). In our calculations we shall compare the heat requirements at a theoretical yield (impossible to obtain and far from the obtainable one) and consider only the heat for change of state; any heat requirement derived from specific heat will not be considered because of its small relative importance.

Working in an evaporator with liquids above a 2% concentration is impossible, problems are also posed by the gelling temperature of the extract and its large non-newtonian viscosity at temperatures close to gelling. All this prevents the thermal savings that could be gained by the use of multiple-effect vacuum evaporators.

B. PRECIPITATION

A working method similar to that used for carrageenan, using alcohol precipitation, could be considered. An economical process using this technique has not been achieved so far but the process is feasible chemically. Agar precipitation in alcohol media is more difficult than for carrageenan because the precipitate is more flocculant (has low cohesion) and is difficult to recover quantitatively. A high heat consumption is required because we have to add the heat needed to evaporate the alcohol, to the 53 361 kcal needed to evaporate the water in the mixture. In addition, the alcohol used for precipitation has to be recovered by distillation for reuse.

If we make our calculations using isopropanol, which is used for producing carrageenan for economic reasons, and consider that we start with an azeotropic mixture, previously recovered, of 87% by weight we are forced to work at least 3 litres of azeotropic isopropanol for each litre of extract to be precipitated. Assuming that such a mixture has a density of 0.8234 kg/L, then for each kilo of agar it would be necessary to evaporate:

        99 kg of water extract;
        22.5 kg of azeotropic isopropanol.

This second item is composed of 213 kg of isopropanol and 41.5 kg of water. The latent heat enthalpy for isopropanol is 175.8 kcal/kg. Therefore the theoretical heat energy consumption would be:

      water: 539 x (99 + 41.5) = 75 729 kcal;
      isopropanol: 175.8 x 213 = 37 445 kcal.

This means an energy consumption of 113 174 kcal/kg of agar which is double the heat energy need to evaporate the water contained in the extract, and all of this without taking into consideration the need to concentrate the used alcohol back to the azeotrope plus the recovery of isopropanol vapors that entail a considerable amount of energy. From this it can be seen that the precipitation/dehydration process analogous to that used for carrageenan has a high energy consumption when applied to agar.

For agar, concentration methods are based on its insolubility when cooled and are used in all factories according to two basic principles: freezing or syneresis under pressure.

C. FREEZING

This consists of freezing and thawing the extract, previously gelled, and profiting from the insolubility of agar in the cold to eliminate the greatest part of the water contained in the extract. Freezing should be slow, to allow both the growth of ice crystals and the separation of agar in the highest possible concentration; this is usually followed by draining with a water-extracting centrifuge. Only slow freezing permits large ice crystals to be formed, surrounded by fine sheets of agar. Efforts to speed up freezing produce spongy masses, with high water content and less agar concentration, that dialyse poorly and produce an impure agar, because the impurities which are soluble in cold water do not move so well from the gel to the water.

As far as the economics for this process are concerned, we should consider that if we start with a 1% agar extract, we have to eliminate 99 litres of water per kg of agar; after melting and draining, this agar at best reaches a dry extract content of 15% (1 kg in 6.66 L) but is normally 11-12%. Presuming a 15% agar in the product, the cycle of freezing-defrosting eliminates (99 L - 6.66 L) 92.34 litres water per kg of agar. Furthermore the energy consumption for freezing 1 L or 1 kg of water is 79.67 kcal. To freeze the 99 litres of water contained in the 1% extract would require:

$$99 \times 79.67 = 7\ 887 \text{ kcal}$$

To remove the water remaining in the melted and drained agar requires a heat consumption of:

$$6.66 \times 5\ 390 = 3\ 590 \text{ kcal}$$

We can see the difference between the sum of these figures (11 477) and those for:

evaporation method = 53 361 kcal,
precipitation method = 113 174 kcal.

Naturally there is a cost difference between obtainig a difference of a kilocalorie by heating or cooling but the figures leave us in no doubt (even though we have ignored the energy consumption derived from the specific heat of the water that is eliminated) there are enormous energy differences between the working methods considered.

D. SYNERESIS

Syneresis is usually des cribed as the process in which a gel contracts on standing and exudes a liquid. Here the term syneresis is used to describe the process where pressure is used to exude liquid from the gel. The water that soaks the colloidal net of the gel is eliminated by applying, by suitable means, a force that will favour such loss. Energy consumption is very low when working in these conditions but not everybody can benefit from it because the industrial technology is not simple. Pressure has to be applied very carefully to avoid gel losses by extruding the gel through the containing system. The advanced factories that use this process have been obliged to develop a very specific technology, not only producing extracts in the appropriate conditions for good syneresis but also equipment design that will allow the efficient treatment of large quantities of extracts.

Initially long syneresis periods were required, with cycles longer than 24 hours, that would start with a gradual and slow increase in pressure by placing, successively and at a prefixed rate, stone blocks on top of the gel containers; the agar gel was wrapped in canvas cloths and placed in a series of steel boxes fitted between the fixed and movable heads of a vertical hydraulic press. This treatment was followed by hydraulic pressing, once the product was consistent enough to withstand extrusion. Usually a modified platen press is used which is similar to a box press but the cloth bags are not enclosed on the sides during pressing and the press is usually built in horizontal form. Nowadays some agar manufacturers have designed their own modern equipment which permits this syneresis to be carried out automatically in relatively short periods of time and operating with large volumes.

Starting with a 1% agar extract, syneresis increases concentration to a maximum of 25% (1 kg agar per 3 L water). If we consider an average of 20% for the dried extract from industrial runs, the heat energy necessary to remove the rest of the water will be:

$$4 \times 539 = 2\ 156 \text{ kcal/kg}$$

which is much less than the heat energy needed to dry the agar obtained by freezing where moisture was calculated in ideal conditions, that are difficult to obtain in reality.

Compared to freezing, syneresis results in large electrical energy savings as the electrical energy needed to maintain a pressure on a quite incompressible product is much less than that necessary to freeze 99 litres of water for each kg of agar produced. The cost of electrical energy makes many freezing factories increase extract concentration but this is possible only up to 1.5% before producing a harmful effect from yield losses. Syneresis, when properly applied, will also produce a purer agar, eliminating a larger quantity of soluble matter.

6. GENERAL

Figure 11 is a flow chart showing the steps used in both of the dehydration processes used to produce agar. Treatment and reagents used in each case will be very variable depending on the species of seaweed used, its origin and even the time of the year when it was harvested. All these factors can cause drastic modifications to the treatment.

Nevertheless, we should consider some general rules. Seaweeds such as <u>Gelidium</u>, <u>Pterocladia</u> and <u>Gelidiela</u> can be treated by different diffusions, the most usual ones being sodium carbonate solutions, at about 80-95°C. Other reagents such as calcium or aluminium hydroxides or salts can also be used for several purposes. Other treatments with sodium hydroxide solutions of very variable concentration can be used, but the concentration will vary depending

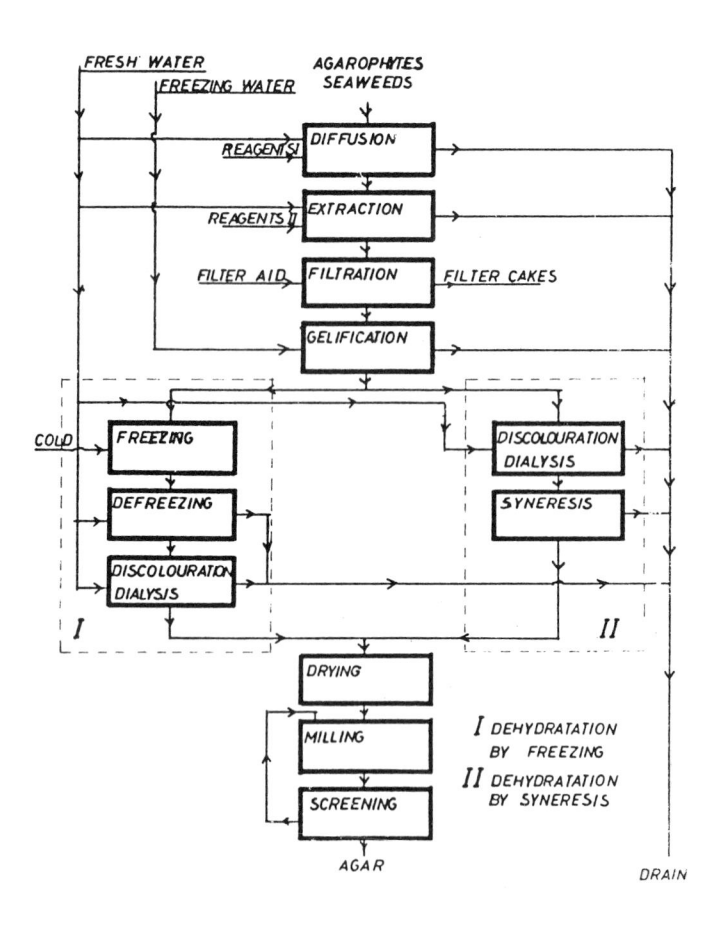

Figure 11   Agar production diagram

on what purposes they are for. As far as <u>Gracilaria</u> is concerned, 0.1 M sodium hydroxide solutions are commonly used; higher concentrations can also be used. The reagents named "Reagents I" in Figure 11 are basically the ones mentioned above. The ones shown as "Reagents II" are those used to adjust the extracting conditions and, in general, are organic or inorganic acids or salts with which pH and other extraction parameters are fixed.

The variables in the manufacturing process make it hard for a factory to change the seaweeds it uses as raw materials. Agar manufacturing history is full of fiascos caused by industries trying to change their seaweeds without having adequate technology to adapt to the change.

Water consumption in an agar factory varies widely depending on the seaweed used but it is always very high. Normally factories working <u>Gracilaria</u> seaweeds have a higher water consumption than others. Consumption also increases when an agar of better quality is required, although, in general, it can be reduced by a suitable design of the factory; however this can lead to an increase in investment and therefore to a more difficult project profitability. Factories using the freezing process have very high water consumption as cooling water is needed for the freezing equipment.

Using recycled water, after appropriate treatment, would reduce its consumption but, in general, would increase the plant operating cost. If poor quality water is going to be used, a prior treatment will be required but it is very important to know its cost before the location is decided since a mistake in this point could make the operation of the factory economically impossible.

The above-mentioned problems about water, and those originating from changing to seaweeds of a different origin, are the ones which have led many factories to bankruptcy.

A manufacturer of good quality agar must be ready to monitor his process and so be able to spot readily any variations that seaweeds cause in the yield or in the quality of the final product. For this purpose a well equipped control laboratory is required together with a pilot plant that will enable any modifications needed in the process to be studied prior to the industrial treatment of each batch of raw material. An adequate pilot plant can process from 1-10 kg of seaweeds, depending on the size and importance of the factory. In general small factories with elementary technology do not achieve international quality standards and their products have to be sold at lower prices in local markets. Bacteriological contamination particularly is usually too high and sometimes dangerous in such plants, closing them to many markets.

A food grade agar should have a moisture content of less than 18%, ash below 5%, gel strength above 750 g/cm^2 (Nikan-Sui method) and a bacterial count below 10 000 bacteria per gram. Escherichia coli and Salmonella must be absent (other pathogenic bacteria may also be specified). Usually the lead content is specified as less than 5 ppm and arsenic less than 3 ppm. These specifications are for agar produced on an industrial scale. In the Orient, large quantities of "natural agar" are sold by very small producers and consumed in the form of threads ("strip") or bars ("square") that are usually produced from Gelidium and do not have to meet the above-mentioned specifications. Generally its gel strength is 450 g/cm^2 by the Nikan-Sui method.

Figure 4b shows as closely as possible what we consider the present situation for the world production of agar. This table has been prepared taking into account the results obtained from an enquiry made among the most important agar manufacturers in countries such as Spain, Chile, Morocco, Portugal, Argentina, Mexico, France, New Zealand, Brazil, etc., and the available Japanese statistics. All these data along with others from Korea, People's Republic of China, its Taiwan Province and Indonesia have been updated during the XIIth International Seaweed Symposium held in Brazil, August 1986.

BACTERIOLOGICAL AGAR

The use of agar in bacteriology is one of the most important uses and requires strict physical-chemical control as well as the absence of hemolytic substances and what is more important and difficult, the absence of any bacterial inhibitors. Robert Koch started using agar in 1881 to gel culture broths when preparing solid culture media and this was the first introduction of this oriental product to Europe.

Its uses in microbiology are based on the special properties: a gelling temperature of 32-36°C, a melting temperature of 85-86°C, a lack of hydrolysis by bacterial exoenzymes and its ability to be prepared without bacterial inhibitors. The above temperatures refer to culture media gelled with agar and which contain 10-11 g agar per litre of culture media.

Bacteriological agar is prepared from Gelidium and Pterocladia because Gracilaria and Gelidiella give agars with gelling temperatures above 41°C. It is manufactured in a limited number of highly specialized factories and under rigid physical-chemical and bacteriological controls.

There are no specifications for a universal application for bacteriological agar as the different microbiological schools evaluate the parameters in various ways. There are neither international nor national specifications. There are many differences between food grade and bacteriological grade, in physico-chemical and bacteriological controls, but this information is confidential and is shared only by the bacteriological agar and culture media manufacturers.

As agar is used only as a gelling agent in solid media, it is essential to avoid interactions with the rest of the media components such as meat extract, peptones, proteins, amino acids, sugars and other carbohydrates, as well as pigments, indicators, inhibitors, mineral salts, etc., used in their formulation. It has to mix with these components without producing problems such as colour changes, precipitate formation or gel strength losses, even after autoclave sterilization. Therefore actual specifications are different depending on each user and each culture media manufacturer. In general, bacteriological agars are very transparent agars in solution as well as in gel form and they represent the purest qualities in the world market. The rest of the parameters vary as the agars are adapted to the individual requirements of the manufacturer and end user.

In much smaller quantities, and at a much higher price, another type of agar called "Purified Agar" is also available. These are bacteriological agars that could also be used in biochemistry for electrophoresis or immunodiffusion; they can be considered as agarose forerunners, being still used for economic reasons.

AGAROSE

The composition of this agar fraction has already been explained in the section dealing with the chemical structures of agar. In the literature we have found that agarose had been prepared according to at least 15 basic principles starting with the acetylation procedure of Araki (1937). A list of these methods follows even though they are interesting mainly for historical reasons.

1. Acetylation. This method is based on the different solubility in chloroform of the acetates of agarose and agaropectin.

2. Selective solution. This is based on solubility differences between agarose (less soluble) and agaropectin (more soluble) in aqueous media in well established conditions.

3. Quaternary ammonium precipitation. This is classical method worked out by Hjerten (1962) and based on the insolubility of products resulting from the reaction of agaropectin with some quaternary ammonium salts.

4. Polyethylene glycol. The classical method of Polson (Russell, Mead and Polson, 1964; Polson, 1965) based on the reduced solubility of agarose in media that contain polyethylene glycol.

5. Dimethyl sulfoxide extraction. Tagawa (1966), the method is based on the different solubilities of agarose and agaropectin in this solvent.

6. Ammonium sulfate precipitation. Azhitskii and Kobozev (1967), the method is based on the precipitation of agaropectin with ammonium sulfate.

7. Ion exchange. Zabin (1969), the method is based on ion exchange in citrate or acetate forms.

8. Insoluble support absorption. Barteling (1969), the method is based on the absorption of agaropectin on a non-reactive support such as aluminium hydroxide gel.

10. Chromatography. Izumi (1970), the method is based on a chromatographic separation of agarose and agaropectin.

11. Acrinol precipitation. Fuse and Gotto (1971).

12. Electrophoresis. Hjerten (1971), the method uses electrophoresis over granulated or non-granulated agar gels or over powdered agar.

13. Rivanol precipitation. Svridov, Berdnikov and Ivanov (1971), the method depends on the precipitation of agaropectin with rivanol.

14. Chitin and chitosan precipitation. Allan et al. (1971), this method uses the absorbent chitin or chitosan to eliminate agaropectin.

15. Ethanol or 2-methoxyethanol precipitation of agarose dissolved in a urea buffer. Patil and Kale (1973).

Based largely on these methods, other publications and patents have appeared modifying or maintaining these principles for processes for the preparation of agarose. Sometimes two or even three fractionating methods have been used successively in attempts to improve the agarose quality. At this time we have records of four companies manufacturing agarose and only one of them is an agar manufacturer, very different equipment is needed for the two kinds of production. For agarose, a quality control laboratory with very sophisticated analytical equipment to analyse the finished product is essential. Continuous improvement in technology is essential to adapt to modern applications in biochemistry which have required the introduction of modifications in the chemical structure of agarose, by synthetic organic chemistry in many cases. Thus, an agarose sample obtained from a manufacturer of biochemical reagents does not correspond normally to what we can extract from agar by any of the methods previously mentioned.

Criteria for judging agarose are multiple and they can be grouped in the following way.

A. Physico-chemical properties. In this case the same basic criteria as for agar are followed: colour, transparency in solution, moisture, ash (in this case much lower due to the absence of polar groups), gel strength, gelling and melting temperatures.

B. Purity critera. Reduction of electronegative groups to the minimum, the effects of such groups include an electroendosmosis increase and also an increase in the fixation of electrically charged substances, such as an increase in non-selective fixation of proteins. The increased presence of electronegative groups can also be produced by poor separation from the agaropectins. Likewise it is very important to assure the absence of residues of reagents used in the agarose production process.

C. Specifications are necessary for practical applications, such as protein electrophoresis, DNA residues, non-selective fixation of proteins. For example the absence of inhibitors that could hinder the DNA recombining fragments split by agarose techniques. Controls that will prove agarose to be acceptable for biochemical techniques are included in this group.

Generally, the first two groups appear in specifications even though in some cases the data offered causes confusion, as happens for example in electroendosmosis. Although an accepted criteria for purity is a low electroendosmosis (less electronegative groups present) there are agaroses that have a greater electroendosmosis and yet are better in some specific biochemical separations. Values given to electroendosmosis vary widely for the same agarose when analytical conditions change, such as buffer pH, ionic strength, protein standards and non-charged molecules as well as other conditions dependent on the equipment such as voltage, operating cycle, refrigeration or electrical contact strips. The growing biochemical applications of agarose imply modifications in its structure to expand its range of uses. Thus, it is not realistic to set detailed specifications for a continuously evolving product and none have been set at a national or international level. Some typical specifications for commercial agarose can be found in the Sigma Catalogue (Sigma Chemical Co. 1987) and FMC offer their analytical methods to scientists in their catalogue, "Marine Colloids 1981 Bioproducts Catalog".

PROPERTIES

The most important characteristcs of agar are the following.

1. Its great gelling power in an aqueous environment allows it to form gels which are more resistant (stronger) than those of any other gel-forming agent, assuming the use of equal concentrations.

2. The simple water solution has that gelling power. There is no need to add reagents to produce gelation, such as potassium (or proteins as is necessary with carrageenans), calcium (or other divalent cations as is necessary with alginates). High sugar concentrations or an acid environment (as is necessary with pectins) are not needed.

3. It can be used over a wide range of pH, from 5 to 8, and in some cases beyond these limits.

4. It withstands thermal treatments very well, even above $100°C$ which allows good sterilization.

5. A 1.5% aqueous solution gels between $32°C-43°C$ and does not melt below $85°C$. This is a unique property of agar, compared to other gelling agents.

6. Agar gives gels without flavour and does not need the additions of cations with strong flavours (potassium or calcium), it can be used without problems to gel food products with soft flavours.

7. It assimilates and enhances flavours of products mixed with it and acts as a fragrance-fixer permitting their long term fixation.

8. Its gel has an excellent reversibility allowing it to be repeatedly gelled and melted without losing any of the original properties.

9. Transparent gels that are easily coloured can be obtained whose refractive index can also be easily increased by adding sugar, glucose, glycerine, etc., given them an attractive brightness.

10. The gel is very stable, not causing precipitates in the presence of certain cations as happens to alginates with calcium.

The gelling properties of agar are the origin of its multiple applications; it has the highest natural gel strength of any gelling agent. Terefore the methods used to measure its gel strength are important, as is a knowledge of the structure of the gel. Both these matters will be discussed.

MEASURING GEL STRENGTH

The following methods for measuring gel strength are basic controls in the agar market yet they are not mentioned in the Pharmacopeias, National Formulary, Codex or other similar publications of specifications and analytical methods referring to agar. This results in confusion as these methods are generally used for food grade agar, bacteriological agar and for agarose, by industry and commerce.

The Nikan-Sui method is the most common one used to measure the agar gel strengh. This method is based on measuring the load (g.cm^{-2}) that causes a standard gel to break in 20 seconds. A hot 1.5% solution is poured into metallic boxes (6 x 30 cm base, 4.5 cm high) to the 3 cm level, leaving it to gel at 20°C. The breaking load withstood for 20 seconds is measured with an apparatus designed by the engineer Takenami and made by KIYA SEISAKUSHO LTD., 50 Komagomo, oiwake-cho, Bunku, Tokyo, Japan, (see Figure 12). The solution is made in such a way that total solution of the agar and a final concentration of 1.5% are guaranteed. Boiling and stirring must be maintained long enough to avoid the agar sticking to the bottom of the box; this is achieved by boiling under reflux or by adding hot water to maintain the initial weight and so keep the concentration constant. The load used is a cylindrical plunger with a frontal area of 1 cm^2.

Another technique used in some markets is based on a **Rowerbal weighing machine** (Figure 13) which adds increasing loads until the gel ruptures. The gel, 1.5% agar, is prepared in a similar way to the one used for the Nikan-Sui method but in a crystallizing dish (70 mm diameter, 52 mm high) to a level of 48 mm. The value obtained by this method differ a little from the ones obtained by the Nikan method, so it is always important to state the method used for gel strength control.

The gel strength of a 1.5% solution of industrial agar lies between 600 and 1 100 g.cm^{-2} (Nikan-Sui method); the strength of the normal quality is 700-800 g.cm^{-2}. This is between five and eight times higher than the gel strength of other colloids used in the food industry.

GEL FORMATION AND STRUCTURE

In agar gels, helicoidal structures have been verified by X-ray diffraction, similar to those found in carrageenan. However because agar contains 3,6-anhydro-L-galactose, the helices are left-handed whereas in kappa and iota carrageenan, which contains 3,6-anhydro-D-galactose, they are right-handed (dextrogyres). Also the helix pitch is shorter than the 26A° of carrageenan. This is explained by several authors (Arnott et al., 1974) as being due to the lower content of sulfate groups that possibly cause a tighter and more compact net. Molecular configuration changes and agarose interaction in sol-gel transitions have been well studied through ultra vacuum circular dichroism (u.c.v.d).

The gelation process from solution (colloidal sol) happens as per the scheme shown in Figure 9 which shows the different steps starting from random coil, through the left-handed dual helix formed by hydrogen bond formation that then will be the base for the macro grid that will give the gel rigidity. The hydrogen bond formation can be obstructed and even prevented if a proton-catching agent (like urea or

Figure 12 Gel strength measurement. Nikan-Sui Method

Figure 13 Gel strength measurement. Rowerbal method

guanidine) is added to the sol to be gelled. Such additions prevent agar or agarose gelation and result in a solution, similar to glycerin, which when cooled does not gel. By removing the proton catcher, the hydrogen bonds will form and therefore the gel-forming ability will be restored. In the same way, considering that dry agar, be it in powder, flake, square or strip from, is really a dry gel (xerogel), its solubility in the cold is not possible as it maintains the hydrogen bonds formed during the gelation prior to its dehydration.

A fundamental characteristic of an agar and agarose gel is what can be called "gelation hysteresis". An agar or agarose gel, when cooled, forms a gel at temperatures between 32° and 43°C depending on the seaweeds used, as that will determine the presence of a variable quantity of methyl groups. However when the well formed gel is heated, a temperature of 85°C must be reached to get the gel to melt and to become a sol. Such a big difference between gelling and melting temperatures is exceptional when compared with the rest of the phycocolloids. It is explained by a greater number of hydrogen bonds and the lack of sulfate groups, which produce a gel with helix pitches much shorter than those of carrageenans, and that, in contrast, does not show cation reactivity.

The characteristic of "viscosity hysteresis", is also remarkable. This can be demonstrated by a solution or colloidal sol prepared at boiling point and held in a thermostatic bath, for example at 80°C, and then its viscosity measured. Afterwards it is held at a lower temperature, at 50°C for example (above the gelling temperature of the sol) keeping it there for a few hours. Subsequently it is held again at 80°C and once this temperature is reached its viscosity is determined again, and it gives values higher than those initially measured. When this temperature is maintained, viscosity values decline slowly almost down to the values measured the first time. Therefore the viscosity values obtained for a solution of agar could depend on its previous history.

USES

Agar was the first phycocolloid to be used in the human food industry. In the beginning it was only used in the Far East, but the applications have been extending all over the world for more than a century. The increasing range of applications is due to the particular gelling characteristics which are not present in any other phycololloid, gum or gelatin. As a result the price for food grade agar is higher than that of other phycocolloids with gelling properties which are also permitted as food additives. In addition, these characteristics allow agar to be used successfully and even exclusively in certain scientific and industrial applications. Some earlier reviews have also discussed uses of agar (Selby and Wynne, 1973; Meer, 1980; Glicksman, 1983).

The 10 important characteristics of **agar**, listed at the beginning of the "Properties" section, explain technically many of the applications of agar. For use by the human food industry its safety is guaranteed by more than three hundred years of non-interrupted use by some countries and for more than a century on a world scale. In addition the FAO/WHO Codex Alimentarius permits the use of agar in the human food industry and it has also been accepted and authorized by the regulations of the more exacting countries such as United Kingdom, Federal Republic of Germany, Russia, France and Poland. The Food and Drug Administration (FDA) of the United States assigns agar as a grading of GRAS (Generally Recognised as Safe).

In the human food industry, agar is used mainly as a gelling agent and in a secondary way as a stabilizing agent and for controlling viscosity. It is used as an additive, not as a nutrient. The gelling power of agar is so high that it is used at 1% maximum concentration; for viscosity control and as a stabilizing agent the proportion used is 1/100 or less. For this reason the ingested quantities are very small and, because agar is not easily digested by the human body, its calorie contribution is negligible and thus agar can be used in special diet food. Agar digestion by the human body is imperfect, studies have shown that less than 10% of the polysaccharide is assimilated. Therefore due to the small proportions in which it is used in human food, its importance as a nutrient is very small.

Agar applications in the food industry are based on its special characteristics and the most important applications are the following.

In confectionery, to prepare jellies, marshmallows and candies or candy fillers.

In marmalade production, agar is used as a thickening and gelling agent.

Mitsumame production in Japan is very important; this is a fruit salad mixed with agar gel cubes, duly coloured, salted and flavoured with fruit flavour. The agar used for this kind of fruit salad must allow the cans to be sterilized without the cubes melting or losing their corners or edges. For this purpose certain types of Gelidium agar are used.

In bakery, agar is used to cover cakes, in icing doughnuts, and when it is applied to chocolate it allows a good adherence to the base without cracking. In general agar is utilized to prevent dehydration of these confectionery products.

Agar is also important in fruit jelly preparations. When compared with pectin, agar has the advantage of not needing high sugar concentrations to form a gel.

Its application in yoghurt is also very important especially when consumers started to require less acid products and, therefore, casein cannot contribute to the maintenance of the product consistency, as it previously did.

In the meat industry, and especially in the preparation of soft boiled sausages, its use has permitted the reduction of fat content that acted before as bonding. Today the industry is trying to limit fat content in order to reduce cholesterol.

Agar is also used on a large scale in canned products like "scatola" meat (beef blocks in gelatine) - very popular in Italy, or chicken in gelatine - very common in Canada, cow tongue in gelatine - selling well in Denmark, lamb tongue in Australia, or other different types of meat and fish aspics. In dressings and extracts it is used as a thickener and stabilizer.

In smaller quantities, agar is used to increase the viscosity of some alcoholic liquers.

The important gel-forming properties of agar have permitted the expansion of its use to applications other than the human food industry. To prepare casting moulds, an aqueous solution of high concentration (8% or more) is used with the addition of glycerin or glycols, as well as a preservative to avoid gel surface contamination by moulds; bacterial contamination is impossible because a gel with such a high concentration has very little free water left where bacteria can grow (it has a very low A_w). These kinds of gels are also used in sculpture, archaeology, and in other works in which a perfect or precise reproduction is essential. For this reason it is also used in dental moulds as it is possible to make better and more precise reproductions, in spite of the fact that casting moulds prepared with agar are more expensive than those prepared with alginate paste. The purpose of adding glycerin in these uses is to avoid cast dehydration since a balance with the outside humidity can be achieved and therefore stable gels can also be obtained which do not appreciably change the gelling and melting points. Glycerin also expedites heat transfer permitting a faster gel melting in a boiling water-bath. The perfect reversibility of an agar gel permits it to be repeatedly used in this application and its low gel point makes it possible to be used as a dental mould.

Another rare application is the preparation of food to feed insects during their larval stages. The breeding of certain insects on a large scale, such as the Mediterranean fruit fly that attacks fruit trees or Pectinopora glosipeii that attacks cotton plantations, has made plague control possible through a type of biological warfare. Large numbers of insects are grown and sexually sterilized by gamma rays. When released they drastically reduce the reproduction of the overall population.

The application of agar in pharmacy as a smooth laxative is well known. Lately it has been used as an excipient in pharmaceutical preparations. In some Western countries agar is used as an antirheumatic since a prolonged treatment has permitted important improvements in patients' health. Agar has been used to stabilize cholesterol solutions.

Due to its low tendency to precipitate in alcoholic media, it is possible to prepare agar gels with alcohol concentrations that will burn when a match is applied to them.

Agar has been used in orchid nurseries for a long time. Improvement in cellular cultivation know-how has brought another important application of agar mainly in the techniques that, starting from meristems, produce perfect and virus-free clones of plants. Agar for these more refined agricultural applications must be especially prepared and strictly controlled to guarantee the absence of inhibitors that prevent cellular cultivation, or make it difficult. This use is increasing since cellular cultivation media for vegetables is being used in industrial quantities because of the current application of these techniques in agriculture.

The 10 basic points previously mentioned are very important as far as Bacteriological Agar and Agarose applications are concerned, as well as those properties essential for the applications of both products. Agarose is produced from both Gelidium and Gracilaria and these two raw materials can give agaroses with different properties which are useful in various applications. Agarose is employed in biochemical technology for protein separation, mainly in analytical laboratories, but has started to appear in industrial applications due to the impressive growth of the genetic engineering industry that produces proteins such as interferon, interleukin and insulin which are sometimes separated with beads made of agarose. Microbiologists employ it in tissue culture as well as in other very sophisticated techniques. Renn (1984) has described some of the applications of agarose.

The uses of agarose can be grouped in the following broad categories.

1. Immunodiffusion and diffusion techniques

2. Electrophoresis of particles carrying electrical charges with direct application for proteins, nucleic acids, polysaccharides that necessarily are charged in conventional electrophoresis as well as reverse electrophoresis, immunoelectrophoresis or electrofocusing.

3. Chromatographic techniques in gel chromatography, ion exchange chromatography (for which the required polar groups will be fixed in the agarose), affinity chromatography or chromatofocusing.

4. In bioengineering as a raw material for beads used in chromatographic columns for separations of proteins, as well as cross-linked beads to which active molecules can be attached which can be recovered afterwards.

5. In microbiology as an excellent base for growing very special cultures, in many cases related to oncological research.

It can be seen that agarose uses are developing as biochemistry and especially bioengineering advance. It constitutes an inert support of natural origin, modifiable by organic synthesis, with the highest known gelling power among natural colloids. Moreover it is of such a great stability that no reagents are needed to preserve it indefinitely, thus avoiding foreign interference with the products meant to be separated.

MARKETING

Agar is marketed pure, without adding mineral salts, other gums or products destined to standardize its properties as is done with carrageenans and some alginates. For that reason, it is very important to know the different techniques of gel strength measurement since the industry produces agars of different gel strengths. The use of agar combinations with other gums for several applications, to solve different rheological problems, makes it necessary to explain here its behaviour when mixed with substances to obtain certain textures. In some applications, agar by itself gives a brittle texture and to improve its elasticity, it is mixed with locust bean gum (also called carob gum) to obtain more elastic gels. In this case Gelidium agar behaves differently from Gracilaria agar as shown in Figure 14 which illustrates Nikan-Sui gel strength of solutions of agar-carob gum mixtures of 1.5% total colloid concentration. It can be seen that for Gelidium agar, the gel strength increased when substituting part of the agar by carob gum, reaching its maximum strength at an approximate concentration of 1.33% of agar and 0.17% of carob gum. Any further increase of carob gum produces a drop in the gel strength. In the case of Gracilaria agar, the addition of carob gum produces a drop in the gel strength. On the other hand Gracilaria agar used in high sugar concentration solutions (above 50%) increases its gel strength much more than Gelidium agar does.

The data for all these applications refers to industrial agars that are sold worldwide in powder form with different meshes, generally included between 60-100 mesh, ASTM standards. in the Orient there is a considerable household consumption of natural agar with much lower gel strengths, in the range 150-400 g/cm^2, with which daily food is prepared at home. This agar is marketed in "strips" or "square" (the appearance is string-like or bar-like respectively, Figure 15) produced always by freezing, thawing, draining, and drying without breaking the strips or squares, that are prepared at the gel stage. This popular method of presentation helps the housewife with her measurements. Such agar, produced basically in Japan, Korea, People's Republic of China and its Taiwan Province, is consumed locally and exported mainly to neighbouring countries with some quantities being sold in Western countries, mainly in health food stores. In Japan the sale of industrial agar for these uses is successfully presented in a pill form of the same content as a bar, to help the housewife with her measurement of it for cooking purposes.

Marketing of industrial agar, food grade, is generally made through trading companies operating from Japan, or in Europe (where the more important commercial centres are located in Hamburg or London), or in the United States where the most important trading companies are located in the areas close to New York.

It is difficult to give an idea of the prices of commercial agar because the usual trade statistics list in the same table, agars with different specifications and applications and therefore with different prices. However taking as a reference the Japanese statistics (since this country produces, imports, and exports a large quantity of the world agar production) we can see in Figure 16 the export/import prices for 1984-86.

From Figure 16 the average prices for 1984 and 1986 were as follows.

Export	1984	1986
Agar, natural, strip	US$ 16.17 per kg	US$ 21.07 per kg
Agar, natural, square	US$ 28.76	US$ 45.61
Agar, industrial, powder	US$ 12.76	US$ 17.30
Import		
Agar, natural, strip	US$ 10.12	US$ 13.38
Agar, industrial, powder	US$ 12.74	US$ 21.40
Exchange rate	Yen 246.07/$US	Yen 154.23/$US

Due to the large quantities of agar traded by Japan, this data is reasonably representative of the agar world market although the 1986 figures would represent the top market values. This is because the recent strengthening of the yen against most of the major world currencies has allowed some other countries to sell more cheaply.

Part of the industrial agar marketing is handled by companies that trade with food additives as pure products (like agar) or by the preparation of mixtures exclusively formulated for each application.

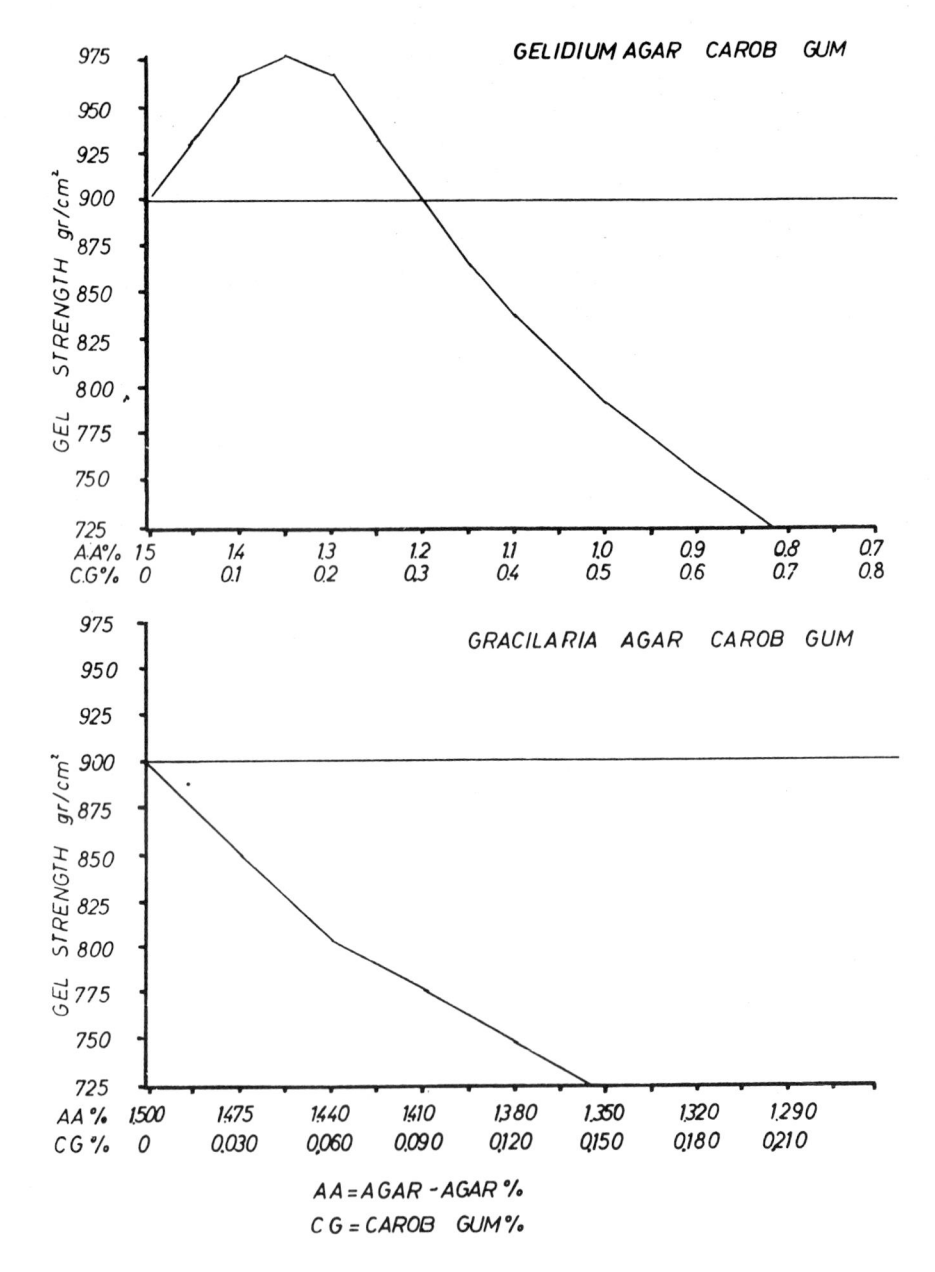

Figure 14 Agar gel solutions of agar/carob gum

Figure 15 Agar in powder, strip and square forms

		----------1984----------		----------1985----------		----------1986 (JAN./OCT.)----------	
		QUANTITIES METRIC TONS	F.O.B. AMOUNT IN YENS	QUANTITIES METRIC TONS	F.O.B. AMOUNT IN YENS	QUANTITIES METRIC TONS	F.O.B. AMOUNT IN YENS
EXPORTS							
I	NATURAL AGAR						
	Strip	7,705	30,666,000	9,186	36,872,000	6,664	21,651,000
	Square	5,779	40,894,000	5,000	35,313,000	3,430	24,127,000
II	INDUSTRIAL AGAR						
	Powder	866,069	2,719,344,000	790,248	2,666,870,000	447,559	1,194,436,000
	TOTALS	879,553	2,790,904,000	804,434	2,739,055,000	457,653	1,240,214,000
IMPORTS							
I	NATURAL AGAR						
	Strip	119,674	298,283,000	65,065	179,054,000	103,078	212,693,000
II	INDUSTRIAL AGAR						
	Powder	331,386	1,039,658,000	420,293	1,356,630,000	227,231	750,114,000
	TOTALS	451,060	1,337,941,000	485,358	1,535,684,000	330,309	962,807,000
EXCHANGE RATE:	1US DOLLAR	------ YENS 246.07 ------		------ YENS 202.60 ------		------ YENS 154.23 ------	

Figure 16 Agar imported and exported by Japan in 1984, 1985 and 1986 (January-October)

The world market for bacteriological agar is small in relation to the food grade market representing 4-5 % of the total agar sales. A constant and direct contact is established between the technical departments of both manufacturer and consumer, who is in general the manufacturer of dehydrated culture media. Prices are higher than food grade for bulk purchases; they range from US$ 16 to US$ 38 since for some secondary media manufacturing companies a very good grade agar is acceptable but for others with higher standards an agar with distinct specifications is absolutely necessary. However when purchased in one pound jars, the differences between food grade and bacteriological grade become very large since marketing and distribution costs have been added. Researchers buy from this source when they have to formulate their own media. Frequently they use the standard formulations that are produced in large quantities under strict controls by culture media manufacturers, in dehydrated or in ready-to use form, in bottles, tubes or Petri dishes.

A similar situation exists for agarose, the increasing development of which requires a continous contact between both the manufacturer's technical experts and the quality control experts of those companies that distribute biochemical reagents, as well as with the researchers who use agarose in very specialized techniques. Its price is the highest of all polysaccharides derived from seaweed but its market is still small, about 0.2% of the agar market. Prices recorded in commercial catalogues are difficult to compare (for example see Sigma Catalogue for 1987 where thirteen grades of agarose are listed, ranging in price from US$ 535 to US$ 5 400 per kg) as they do not mention the degree to which the corresponding agarose has been modified by synthesis and therefore a range of different products are listed under similar prices categories. So unless a good contact is established with the supplier, purchasing agarose can be confusing for the end user.

There is a small demand for Purified Agar, a special grade of bacteriological agar which can be used in some biochemical applications. Its price lies between US$ 38 and US$ 60/kg, the range being explained by the differences of quality in purified, noble or deionized agars present in the market.

REFERENCES

Alkahane, T. and S. Izumi, 1976. Sulfate groups of the mucilage of red seaweed. Agric.Biol.Chem., 40:285-9

Allan, G.G., et al., 1971. A new procedure for the fractionation of agar. Carbohydr.Res., 17:234-6

Araki, C., 1937. Acetylation of agar-like substance of Gelidium mansii L. J.Chem.Soc.Japan, 58:1338-50

Arnott, S., et al., 1974. The agarose double helix and its function in agarose gel structure. J.Mol.Biol., 90:269-84

Azhitskii, G.Y. and G.C. Kobozev, 1967. Ammonium sulfate use to eliminate first agaropectin and then to precipitate agarose. Lab.Delo, 3:143-5

Barteling, S.J., 1969. A simple method for the preparation of agarose. Clin.Chem., 15:1002-5

Bhattacharjee, S.S., G.K. Hamer and W. Yaphe, 1979. Study of agar and carrageenan by ^{13}C n.m.r. spectroscopy. Proc.Int.Seaweed Symp., 9:379-85

Chueh, C.T. and C.C. Chen, 1982. Seaweed economics in Taiwan. In Proceedings of the cooperative science seminar on cultivation and utilization of economic algae, June, 1978, Guam, University of Guam Marine Laboratory, pp. 9-16, edited by R.T. Tsuda and Y.M. Chiang

Corongiu, G., S.L. Fornili and E. Clementi, 1983. Hydration of agarose double helix: a Monte Carlo simulation. Int.J. Quantum Chem.Quantum Biol.Symp., 10:277-91

Cristiaen, D. and M. Bodard, 1983. Spectroscopie infrarouge de films d'agar de Gracilaria verrucosa (Huds) papefuss. Bot.Mar., 26:425-7

Cross, A.D., 1964. An introduction to practical infrared spectroscopy. London, Butterworth

Di Ninno, V. and E.L. McCandless, 1978. The chemistry and immunochemistry of carrageenan from Eucheuma and related algal species. Carbohydr.Res., 66:85-93

_____, 1978a. The immunochemistry of lambda-type carrageenan from certain red algae. Carbohydr.Res., 67:235-41

Fuse, T. and F. Goto, 1971. Some properties of agarose and agaropectin isolated from various mucilaginous substances of red seaweeds. Agric.Biol.Chem., 35:799-804

Glickman, S.A. and I.G. Shubtosova, 1957. Physical chemistry of agar theory and practice of agar fractionation. Kolloid Zh., 19:281-6

Glicksman, M., 1983. Red seaweed extracts (agar, carrageenans, furcelleran). In Food hydrocolloids, edited by M. Glicksman. Baton Raton, Florida, CRC Press, pp. 73-113

Hands, S. and S. Peats, 1938. Isolation of an anhydro-L-galactose derivative from agar-agar. Nature, Lond., 142:797

Hirase, S., 1957. Pyruvic acid as a constituent of agar-agar. Mem. Fac.Ind.Arts Kyoto Tech.Univ.Sci.Technol., 1957:17-29

Hirase, S., C.H. Araki and K. Arai, 1968. The synthesis of agarobiose. Bull.Chem.Soc.Japan, 41:626-8

Hjerten, S., 1962. A new method for preparation of agarose for gel electrophoresis. Biochim.Biophys.Acta, 62:445-9

_____, 1971. Some new methods for the preparation of agarose. J.Chromatog., 61:73-80

Izumi, K., 1970. A new method for fractionation of agar. Agric.Biol. Chem., 34:1739-40

Jones, W.G.M. and S. Peats, 1942. The constitution of agar. J.Chem. Soc., 1942:225-31

Lahaye, M., C. Rochas and W. Yaphe, 1986. A new procedure for determining the heterogeneity of agar polymers in the cell walls of Gracilaria species. Can.J.Bot., 64:579-85

Letherby, M. and D. Young, 1981. The gelation of agarose. J.Chem. Soc.Faraday Trans.1., 77:1953-66

Liang, J.N., et al., 1979. Spectroscopic origin of conformation-sensitive contributions to polysaccharide optical activity. Vacuum-ultraviolet circular dichroism of agarose. Biopolymers, 18:327-33

Meer, W., 1980. In Handbook of water soluble gums and resins, edited by R.L. Davidson. New York, McGraw-Hill, pp. 7.1 to 7.19

Okazaki, A., 1971. Seaweeds and their uses in Japan. Tokyo, Tokai University Press, pp. 98-161

Patil, N.B. and N.R. Kale, 1973. A simple procedure for the preparation of agarose for gel electrophoresis. Indian J.Biochem.Biophys., 10:160-3

Percival, E.G.V., J.C. Somerville and I.A. Forbes, 1938. Isolation of an anhydro-sugar derivative from agar. Nature, Lond., 142:797-8

Pizarro, A. and H. Barrales, 1986. Field assessment of two methods of planting the agar-containing seaweed, Gracilaria, in northern Chile. Aquaculture, 59:31-44

Rolson, A., 1965. Fractionation of mixtures of agarose and agaropectin. British Patent 1,006,259.

Ren, G.Z. and M.Q. Chen, 1986. The effect of temperature on the growth and development of Gracilaria asiatica. Oceanol. Limnol.Sin., 17:292-300

Ren, G.Z., J.C. Wang and M.Q. Chen, 1984. Cultivation of Gracilaria by means of low rafts. Hydrobiologia, 116/117:72-6

Renn, D.W. 1984. Agar and agarose: indispensable partners in biotechnology. Ind.Eng.Chem.Prod.Res.Dev., 23:17-21

Russell, B., T.H. Mead and A. Polson, 1964. Method of making agarose. Biochim.Biophys.Acta, 96:169-74

Samec, M. and V. Isajevic, 1922. Studien uber Pllanzenkolloide.14. Kolloidchemische, Beihefte 16:5-12

Santelices, B. and R. Ugarte, Production of Chilean Gracilaria: problems and perspectives. In Proceedings of the twelfth International Seaweed Symposium. Sao Paulo, Brazil, August, 1986 (in press).

Selby, H.H., 1954. Agar since 1943. Adv.Chem.Ser.Am.Chem.Soc., 11:16-9

Selby, H.H. and W.H. Wynne, 1973. Agar. In Industrial gums, edited by R.L. Whistler. New York, Academic Press, pp. 29-48

Sigma Chemical Co., 1987. Catalog of biochemical and organic compounds for research and diagnostic chemical reagents. St. Louis, Sigma Chemical Co., pp. 169-70

Stancioff, S.J. and N.F. Stanley, 1969. Infrared and chemical studies on algae polysaccharides. Proc.Int.Seaweed Symp., 6:595-609

Stanley, N.F., 1963. Process for treating a polysaccharide of seaweeds of the Gigartinaceae and Soliesiaceae families. U.S. Patent 3,094,517

Sviridov, S.M., V.A. Berdnikov and V.N. Ivanov, 1971. Agarose isolation from agar. Lab.Delo, 1971:55-7

Tagawa, S., 1966. Separation of agar-agar by dimethyl sulfoxide into agarose and agaropectin. J.Shimonoseki Univ.Fish., 14:165-71

Tagawa, S. and Y. Kojima, 1972. The alkal treatment of the mucilage of Gracilaria verucosa. Proc.Int.Seaweed Symp., 7:447-50

Torres-Pombo, J., 1972. Contribución al conocimiento químico del alga Gelidium sesquipedale (clem) Thuret y a la estructura de su agar. Acta Cient.Compostelana, 9:53-64

Tseng, C.K., 1946. Phycocolloids: useful seaweed polysaccharides. In Colloid chemistry: theoretical and applied, edited by J. Alexander. New York, Reinhold, Vol.6:629-734

Watase, M. and R. Nishinari, 1983. Rheological properties of agarose gels with different molecular weights. Rheol.Acta, 22:580-7

Yang, S.S., 1982. Seasonal variation of the quality of agar-agar produced in Taiwan. In Proceedings of the Cooperative science seminar on cultivation and utilization of economic algae, June, 1978, edited by R.T. Tsuda and Y.M. Chiang. Guam, University of Guam Marine Laboratory, pp. 65-80

Yaphe, W., 1984. Properties of Gracilaria agars. Hydrobiologia, 116/117:171-86

Zabin, B.A., 1969. Agarose use of DEAE cellulose to remove the anionic polysaccharides from agar. U.S. Patent 3,423,396

BIBLIOGRAPHY

Blanschard, J.M.V. and J.R. Mitchell, 1979. Polysaccharides in food. London, Butterworths

Determan, H., 1969. Chromatographie sur gel. Paris, Masson et Cie

Glicksman, M. (ed.), 1981. Food Hydrocolloids. Vol.1. Comparative properties of hydrocolloids. Baton Raton, CRC Press

_____, 1983. Food hydrocolloids. Vol.2. Natural plant exudates - seaweed extracts. Baton Raton, CRC Press

International Trade Centre (ITC), 1981. Pilot survey of the world seaweed industry and trade. Geneva, UNCTAD/GATT, International Trade Centre

Levring, T., H.A. Hoppe and O.J. Schmid, 1969. Marine algae. A survey of research and utilization. Bot.Mar.Handb., (1):421 p.

Michanek, G., 1975. Seaweed resources of the ocean. FAO Fish.Tech.Pap., (138)

Percival, E. and R.H. McDowell, 1967. Chemistry and enzymology of marine algal polysaccharides. London, Academic Press

Wieme, R.J., 1965. Agar gel electrophoresis. Amsterdam, Elsevier Publishing Company

INTERNATIONAL SEAWEED SYMPOSIUM PROCEEDINGS

I   Black, W.A.P., et al. (eds), 1953. Proceedings of the First international seaweed symposium. Edinburgh, 14-17 July 1952. Inveresk, Midlothian, Scotland, Institute of Seaweed Research for the Organizing Committee, 129 p.

II Braaud, T. and N.A. Sørensen (eds), 1956. Second international seaweed symposium. Trondheim, 14-17 July 1955. London, Pergamon Press, 220 p.

III Heocha, C.O. (ed.), 1958. Galway, Ireland

IV Davy de Virville, A. and J. Feldmann (eds), 1964. Proceedings of the Fourth international seaweed symposium. Biarritz, 1961. London, Pergamon Press, 467 p.

V   Young, E.G. and J.L. McLachlan (eds), 1966. Proceedings of the Fifth international seaweed symposium. Halifax, Canada, 25-28 August 1965. Oxford, Pergamon Press, 424 p.

VI Margalef, R. (ed.), 1969. Proceedings of the Sixth international seaweed symposium. Santiago de Compostela, Spain, 9-13 September 1968. Madrid, Subsecretaría de la Marina Mercante, Dirección General de Pesca Marítima, 782 p.

VII Nisizawa, K., et al. (eds), 1972. Proceedings of the Seventh international seaweed symposium. Sapporo, Japan, 8-12 August 1971. Tokyo, University of Tokyo Press, 647 p.

VIII Fogg, G.E. and W.E. Jones (eds), 1981. Proceedings of the Eighth international seaweed symposium. Bangor, North Wales, 18-23 August 1974. Menai Bridge, Wales, U.K., Marine Science Laboratories, 769 p.

IX Jensen, A. and J.R. Stein (eds), 1979. Proceedings of the Ninth international seaweed symposium, Santa Barbara, California, 20-27 August 1977. Princeton, N.J., Science Press, 634 p.

X   Levring, T. (ed.), 1981. Proceedings of the Eleventh seaweed symposium . Göteborg, Sweden, August 11-15, 1980. Berlin, Walter de Gruyter, 780 p.

XI Bird, C.J. and M.A. Ragan (eds), 1984. Proceedings of the Twelfth international seaweed symposium. Quingdao, People's Republic of China, 19-25 June 1983. Dordrecht, Netherlands, Dr W. Junk Publishers, Hydrobiologia, Vol.116/117:624 p.

CHAPTER 2

PRODUCTION, PROPERTIES AND USES OF ALGINATES

by

Dennis J. McHugh
Department of Chemistry, University College
University of New South Wales
Australian Defence Force Academy
Campbell, ACT 2600, Australia

SOURCES

Most of the large brown seaweeds are potential sources of alginate. The properties of the alginate varies from one species to another, so the choice of which seaweeds to harvest is based on both the availability of particular species and the properties of the alginate that they contain. The main commercial sources are species of Ascophyllum, Durvillaea, Ecklonia, Laminaria, Lessonia, Macrocystis, Sargassum and Turbinaria. Of these the most important are Laminaria, Macrocystis and Ascophyllum.

Macrocystis is harvested on the west coast of North America, from the Monterey peninsula in central California to the middle of the west coast of Baja California. It has been estimated that the USA harvests about 150 000 t (wet) and Mexico about 40 000 t (wet) per year (ITC, 1981). Nereocystis grows north of the Macrocystis beds but the two overlap and some Nereocystis is also harvested incidentally. Macrocystis was harvested on the east coast of Tasmania, Australia, from 1964-73, but the quantities available were insufficient to sustain an alginate industry.

Laminaria species are harvested principally in Scotland, Ireland, Norway, France, China, Japan and Korea. However in the Asian countries Laminaria is very popular as a food and in Japan and Korea the resulting higher price makes it an expensive raw material for alginate production. In Japan, only the material that is unsuitable for food is used for alginate extraction; since this is insufficient to sustain the alginate industry, other sources have to be found. The situation in China is different; here the cultivation of Laminaria japonica has been very successful, reaching about one million tonnes of wet seaweed annually. About two-thirds of this is used as food and the surplus is available for alginate production; while the cost of cultivated Laminaria is higher than the harvested wild material, the Chinese are able to absorb these higher costs and use the cultivated product for alginate extraction.

Laminaria hyperborea grows on rocky seabeds, usually at depths from 2-15 m, and the upright habit of the plant leads to the use of the phrase, "forests of hyperborea". Stipes that have been cast up by winter storms are collected in France, Ireland, Norway and Scotland. The Norwegian alginate producer, Protan A/S, has developed its own method of harvesting this plant, trawling it with specially built boats that are easily manoeuvred.

Laminaria digitata is found on either side of the low water mark and is usually harvested by hand when the plants are exposed at low tide. It is collected in France, Norway and Scotland but the quantities are small in comparison with Laminaria hyperborea. In France it is harvested using a small boat and a hydraulic arm fitted with a hook device at the end. This is lowered into the bed of Laminaria digitata and rotated so that the weed wraps around it. The arm is then raised to the surface, bringing the seaweed with it.

Ascophyllum nodosum grows in the intertidal zone. It has been harvested by hand in Scotland and Ireland for more than a century. Various attempts have been made to mechanize harvesting but the most successful appears to be that developed by Protan in Norway. This is a nozzle with cutters inside that cut and pump the seaweed through a large diameter pipe into a net bag on a shallow-draught water jet-propelled vessel; the operation is carried out at high tide and the bags can be left floating for later collection. It is also harvested in the southern parts of Nova Scotia.

Durvillaea Lessonia and Ecklonia are used to a lesser extent. Durvillaea antarctica from Chile and Durvillaea potatorum from Australia are used by alginates producers in the UK and USA. In 1985 Chile exported about 390 t and the current exports from Australia are about 3 000 t per annum.

Lessonia is collected in Chile where it is cast up after storms; in 1978 Chile exported 2 045 t of which 1 313 t went to Japan and the remainder to the USA and Canada (ITC, 1981). By 1985 the total export had increased to 5 810 t but no details of countries of destination are available. The North American alginate producers use it to supplement the supply of Macrocystis; the Japanese industry relies mainly on imported seaweeds so Lessonia is one of the primary raw materials.

Ecklonia cava grows in deep water (up to 20 m) and is harvested by divers in both Japan and Korea. Eisenia bicyclis grows in a similar location and is collected along with the Ecklonia in Japan. In Japan, divers find it more profitable to collect the higher priced red seaweeds so the quantities of Ecklonia available are fairly small. Ecklonia that has been cast up by storms is collected in Korea and South Africa; in Korea it is used by the local alginate producer, the South African material is all exported. The Korean industry also uses waste Undaria that is unsuitable for food uses, just as the Japanese industry uses similar waste from Laminaria species.

The alginate obtained from Sargassum and Turbinaria frequently has a poor viscosity so these species are used only when the above colder water species are not available. However recent reports (Shyamali, de Silva and Kumar, 1984; Wedlock, Fasihuddin and Phillips, 1986) on the structure of alginates from warm-water Sargassum and Turbinaria indicate they could be very useful in applications requiring the formation of strong gels (they have a low M/G ratio, see next section). The Indian industry is based on Sargassum that grows in the south (the coasts of Kerala and Tamil Nadu states); the species which grow in the north (Gujarat state) gives a low viscosity alginate, unsuitable for the main Indian market of textile printing; Turbinaria is used only when supplies of Sargassum are unavailable. The Philippines has large resources of Sargassum but this is exported mainly to Japan for use in animal feeds and fertilizers.

PRICES OF SEAWEEDS

There is limited trading in brown seaweeds because many alginate producers harvest the raw material themselves (e.g., Macrocystis, Laminaria hyperborea, Laminaria digitata, Ascophyllum). The dried seaweeds that are traded vary in moisture content and alginate content and this will be reflected in prices. For example air-dried stipes of Laminaria hyperborea contain about 35% moisture while air-dried samples of Lessonia, Durvillaea and Laminaria japonica can vary from 15-20% moisture. A dried tonne of Lessonia could be expected to yield about 140 kg of alginate; some corresponding figures for other seaweeds are, Ascophyllum 120 kg; Laminaria japonica 170 kg; Durvillaea 240 kg. The following are some examples of f.o.b. prices for air-dried seaweed per tonne: Chilean Lessonia US$ 150; South African Ecklonia US$ 250; Australian Durvillaea US$ 400; Chinese Laminaria japonica US$ 500-700; UK Ascophyllum US$ 350. To allow a comparison of the relative costs of these seaweeds, the prices have all be converted to US dollars at the exchange rates current in June 1987. However these prices can show considerable variation from year to year with the fluctuation of the exchange rate of US dollars versus the currency of country of origin of the seaweed.

STRUCTURE OF ALGINIC ACID

Alginic acid is a linear polymer based on two monomeric units, β-D-mannuronic acid and α-L-guluronic acid. The classical Haworth formulas for these monomers are shown in Figure 1, while Figure 2 illustrates the chair formulas, which give a clearer picture of the three-dimensional arrangement of the molecules.

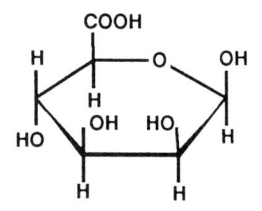

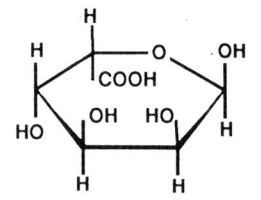

β - D - Mannuronic Acid

α - L - Guluronic Acid

Figure 1 Classical formulas of the two monomeric units of alginic acid

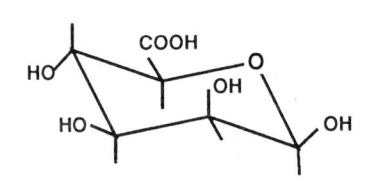

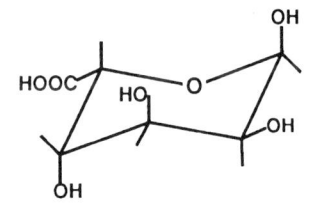

β - D - Mannuronic Acid

α - L - Guluronic Acid

Figure 2 Formulas in Figure 1 expressed as chair forms

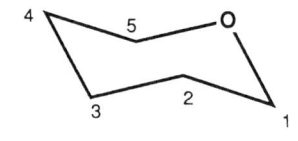

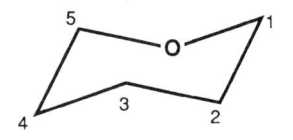

C1 conformation

1C conformation

Figure 3 C1 and 1C forms of the tetrahydropyran ring

The basic structure of each monomer is the tetrahydropyran ring and this has two possible chair forms, C1 and 1C (Figure 3). β-D-mannuronic acid assumes the C1 form; in the other form, 1C, there would be steric interaction between the axial -COOH on C-5 and the axial -OH on C-3; the C1 form has these groups in the equatorial positions and so is more stable. For similar reasons, α-L-guluronic acid assumes the 1C form rather than the C1 form (Penman and Sanderson, 1972; Atkins et al., 1973, 1973a).

The alginate polymer is formed by joining these monomers at the C-1 and C-4 positions. An ether-oxygen bridge joins the carbon at the 1-position in one molecule to the 4-position of another molecule. It has been shown that the polymer chain is made up of three kinds of regions or blocks. The G blocks contain only units derived from L-guluronic acid (Figure 4), the M blocks are based entirely on D-mannuronic acid (Figure 5) and the MG blocks consist of alternating units from D-mannuronic acid and L-guluronic acid (Haug, Larsen and Smidsrod, 1966, 1974; Grasdalen, Larsen and Smidsrod, 1981).

Note the differing shapes of the M blocks and G blocks. Because an M block is formed from equatorial groups at C-1 and C-4, it is a relatively straight polymer, like a flat ribbon. However the G block is formed from axial groups at both C-1 and C-4 so the resulting chain is buckled; the importance of this buckled shape will be apparent later when the formation of gels from alginate solutions is discussed.

So an alginate molecule can be regarded as a block copolymer containing M, G, and MG blocks, the proportion of these blocks varying with the seaweed source. However Larsen (1981) warns that this is an idealized structure which is at best an approximation of the actual situation.

It has been shown that the physical properties of alginates depend on the relative proportion of the three types of blocks (Haug, Larsen and Smidsrod, 1967; Penman and Sanderson, 1972; Smidsrod and Haug, 1972; Smidsrod, Haug and Whittington, 1972). For example formation of gels, by addition of calcium ions, involves the G blocks so the higher the proportion of these, the greater the gel strength; solubility of alginate in acid depends on the proportion of MG blocks present. The industrial utilization of any particular alginate will depend on its properties and therefore on its uronic acid composition so it has become important to have some measure of the relative proportions of the uronic acids. Various methods have been developed to measure the ratio of mannuronic acid to guluronic acid (the M/G ratio) in a sample of alginic acid (Annison, Cheetham and Couperwhite, 1983, and twelve references cited therein). Some examples of M/G ratios are shown in Table 1.

Even more useful, but more difficult to obtain, is a measure of the M, G and MG blocks in a sample and methods have been developed to

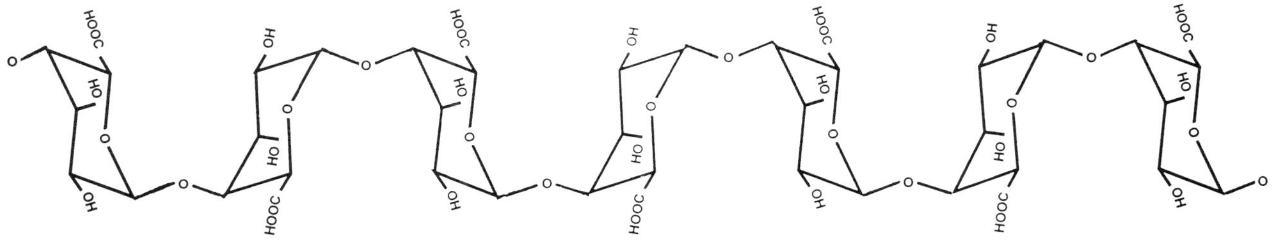

Figure 4 G Block

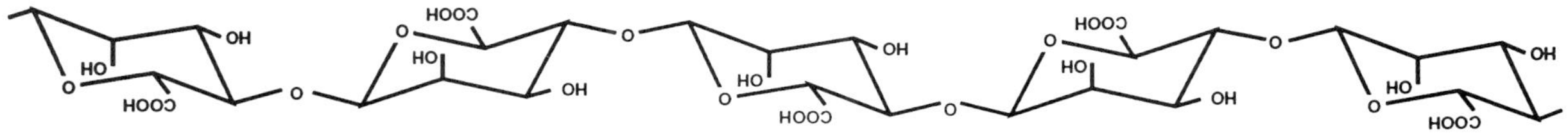

Figure 5 M Block

Table 1

Percentages of mannuronic acid and guluronic acid, and M/G ratios, of alginic acid from various commercial brown seaweeds[a]

		Mannuronic acid (%)	Guluronic acid (%)	M/G ratio
Ascophyllum nodosum	(1)			1.56
	(2)			1.29
	(3)	64.5	35.5	1.82
	(3)			1.10[b]
	(4)	60.0	40.0	1.5
	(5)			1.85[b]
Ecklonia cava, fronds	(6)			2.64-3.08[c]
stipes	(6)			1.39-2.91[c]
Laminaria digitata	(1)			1.45
	(4)			1.63
	(3)	53.7	46.3	1.16
	(3)			1.58
	(3)	59.0	41.0	1.43[b]
Laminaria hyperborea	(2)	38.3	61.7	0.62
fronds	(1)			1.35
fronds	(3)	56.0	44.0	1.28
stipes	(1)			0.65
stipes	(2)			0.40
stipes	(4)	30.0	70.0	0.43
stipes	(3)			0.37
stipes	(3)			0.46[b]
Laminaria japonica	(7)	69.3	30.7	2.26[b]
basal part	(7)			2.34-3.18[c]
apical part	(7)			1.61-2.02[c]
Macrocystis pyrifera,				
Australian	(2)			1.38
American	(3)	61.0	39.0	1.56
frond	(7)			1.52
stipe	(7)	50.5	49.5	1.02
air bladder	(7)			1.41
Undaria pinnatifida	(7)			1.45-2.65[c]

a) Unless otherwise stated, the alginic acid samples were prepared in the laboratory from the appropriate seaweed.
b) A sample of alginic acid made from a commercial alginate.
c) The range shows seasonal variation during one year.

(1) Haug, Larsen and Smidsrod, 1974; (2) Penman and Sanderson, 1972;
(3) Haug and Larsen, 1962; (4) Grasdalen, Larsen and Smidsrod, 1979;
(5) Haug, 1964, p. 108; (6) Kim, 1984; (7) Ji, et al., 1984

achieve this (Haug, Larsen and Smidsrod, 1966, 1974; Penman and Sanderson, 1972; Grasdalen, Larsen and Smidsrod, 1979; Morris, Rees and Thom, 1980). Some examples are shown in Table 2.

The M/G ratio of alginate has been altered, on a laboratory scale, by treating it with "mannuronan C-5 epimerase", and enzyme isolated from the soil bacterium, Azotobacter vinelandii. This enzyme converts mannuronic acid residues into guluronic acid residues in the polymer chain, and the resulting alginate forms stronger gels (Larsen and Haug, 1971; Skjak-Braek, 1984). The method has not been applied on an industrial scale.

The alginate of greatest industrial importance is the sodium salt. Uses are also found for the potassium, ammonium and calcium salts, as well as alginic acid itself. The only synthetic derivative of alginic acid to find wide use, and acceptance as a food additive, is propylene glycol alginate. This is formed by reacting propylene oxide with moist alginic acid (Steiner, 1947; Steiner and McNeely, 1950; Kelco 1952; Pettitt and Noto, 1973; Noto and Pettitt, 1980). Esterification occurs at the carboxylic acid groups on the alginate chain, mainly with the primary hydroxyl group of propylene glycol. Depending on reaction conditions, such as reaction temperature and ratios of propylene oxide to alginic acid, varying degrees of esterification can be achieved. A product with about 60-70% esterification is satisfactory for most purposes but up to about 90% esterification can be achieved and this type of product (80-90%) is useful in very acidic, short term applications.

EXTRACTION PROCESSES

INTRODUCTION

Alginic acid was first discovered by Stanford (1881). An excellent history of the evolution of the alginate industry has been written by Booth (1975). He traces a path from Stanford's successful exploitation of crude extracts to the failure by F.C. Thornley, in Orkney about 1923, to establish a briquette business based on using alginate as a binder for anthracite dust. Thornley moved to San Diego and by 1927 his company was producing alginate for use in sealing cans. After some difficulties the company changed its name to Kelp Products Corp. and in 1929 it was reorganized as Kelco Company. Production in the United Kingdom was established in the period 1934-1939 and in Norway after World War II. It is estimated that there are 17 factories in 9 different countries (ITC, 1981), excluding the People's Republic of China. The two largest producers, Kelco Company in USA and Alginate Industries Ltd in UK, have been acquired by Merck and Co. Inc., USA; these combined companies produce about 70% of the world's alginate. The next largest producer is Protan A/S of Norway, followed by companies in Japan and France (ITC, 1981). Production in China is increasing and is now 7 000-8 000 tonnes per annum.

Table 2

Percentages of the three principal types of block structures in alginic acid, prepared from various commercial brown seaweeds

Alginate from		Polymannuronic segments (M-type,%)	Polyguluronic segments (G-type,%)	Mixed segments (MG-type,%)
Ascophyllum nodosum	(1)	35.0	13.0	52.0
	(2)	38.4	20.7	41.0
	(3)	37.8	21.4	40.8
	(4)	40.0	20.0	40.0
Laminaria digitata	(1)	43.0	23.0	34.0
	(4)	49.0	25.0	26.0
Laminaria hyperborea	(2)	20.3	49.3	30.4
	(3)	23.1	43.3	33.7
fronds	(1)	43.0	31.0	26.0
stipes	(1)	15.0	60.0	25.0
stipes	(2)	18.7	58.6	22.7
stipes	(3)	22.0	64.2	13.8
stipes	(4)	26.0	66.0	8.0
Laminaria japonica	(5)	36.0	14.0	50.0
Macrocystis pyrifera	(2)	40.6	17.7	41.7
	(3)	36.5	18.5	45.0

(1) Haug, Larsen and Smidsrod, 1974; (2) Penman and Sanderson, 1972; (3) Morris, Rees and Thom, 1980; (4) Grasdalen, Larsen and Smidsrod, 1979; (5) Ji, et al., 1984

Some of the early patents still provide useful basic information about alginate extraction (Thornley and Walsh, 1931; Clark and Green, 1936; Green, 1936; Le Gloahec and Herter, 1938; Le Gloahec, 1939) as does work published by the former Institutes of Seaweed Research in Scotland and Norway (Black and Woodward, 1954; Haug, 1964) and more recently by Braud, Debroise, and Pérez (1977). Processes used in Japan have been described by Okazaki (1971).

The minimal requirements for the profitable operation of an alginate extraction plant have been estimated by Moss and Doty (1987). They discuss the minimal seaweed input, colloid output and capital investment needed; they also list estimates of production costs. This analysis is made for agar and carrageenan as well as alginate.

PROCESSES

The chemistry of the processes used to make sodium alginate from brown seaweeds is relatively simple. The difficulties of the processes arise from the physical separations which are required, such as the need to filter slimy residues from viscous solutions or to separate gelatinous precipitates which hold large amounts of liquid within their structure and which resist both filtration and centrifugation.

Processes for the manufacture of sodium alginate from brown seaweeds fall into two categories. Figure 6 is a diagram of the processes, simplified to show their essential difference. In one, the principal intermediates are calcium alginate and alginic acid. In the other, no calcium alginate is formed, only alginic acid.

The advantage of the first process is that calcium alginate can be precipitated in a fibrous form which can be readily separated; it can then be converted into alginic acid which is still fibrous and can also be readily separated. A further advantage of this process is that some calcium alginate can be allowed to remain in the final sodium alginate produced. This gives the manufacturer another method of controlling the viscosity of the final product, as discussed later in the "Properties" section.

The second process does save one step, the formation of calcium alginate, but it also has some disadvantages. When alginic acid is precipitated in this process, it forms a gelatinous precipitate which is very difficult to separate and the overall losses of alginic acid are generally greater than in the former process. The removal of liquid ("dewatering") from within the gel structure of the separated alginic acid also presents difficulties in this second process. The water content in the dewatered alginic acid is often high, so that alcohol must be used as a solvent for the conversion to sodium alginate. This usually makes the process more expensive unless the alcohol recovery rate is very good, and this is not easy to achieve.

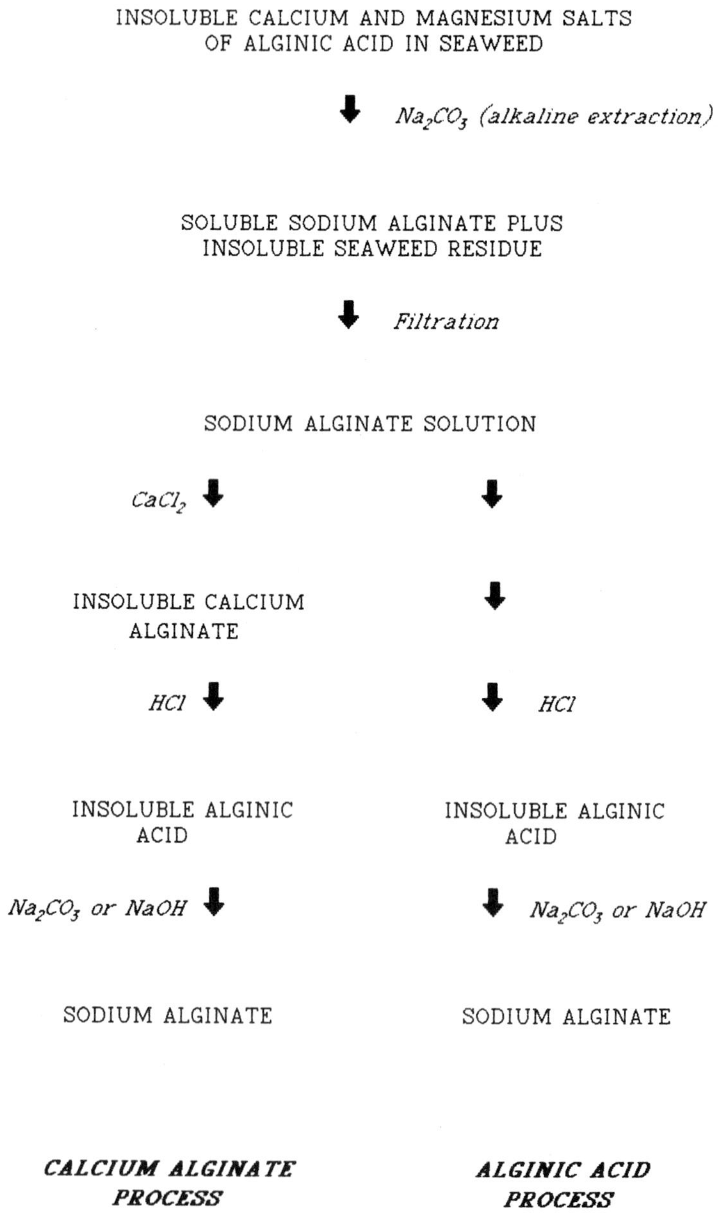

INSOLUBLE CALCIUM AND MAGNESIUM SALTS
OF ALGINIC ACID IN SEAWEED

⬇ Na_2CO_3 *(alkaline extraction)*

SOLUBLE SODIUM ALGINATE PLUS
INSOLUBLE SEAWEED RESIDUE

⬇ *Filtration*

SODIUM ALGINATE SOLUTION

$CaCl_2$ ⬇   ⬇

INSOLUBLE CALCIUM   ⬇
ALGINATE

HCl ⬇   ⬇ *HCl*

INSOLUBLE ALGINIC   INSOLUBLE ALGINIC
ACID   ACID

Na_2CO_3 *or NaOH* ⬇   ⬇ Na_2CO_3 *or NaOH*

SODIUM ALGINATE   SODIUM ALGINATE

CALCIUM ALGINATE   **ALGINIC ACID**
PROCESS   **PROCESS**

Figure 6   Production of sodium alginate

For each process the principles, methods and problems are discussed below.

CALCIUM ALGINATE PROCESS

1. SIZE REDUCTION OF RAW MATERIAL

The raw material may be fresh, dried, or from silage. The last is seaweed which, when fresh, is chopped into small pieces and treated with a dilute formalin solution; this wet weed can be stored in cool concrete containers for several months. Dried weed is rehydrated by soaking for several hours. Several producers do not break up the seaweed into smaller pieces. However, reduction to small pieces, preferably 5-10 mm square, has two advantages. The first is that in the following treatments with formalin, acid and alkali, these reagents will obviously penetrate the seaweed more thoroughly and more rapidly if it has been broken up in this way. The second advantage is that the seaweed can be transported much more readily, by pumping it as a .rry in water.

The size reduction can be done in two stages, the first using equipment to chop the weed into pieces about 20 mm square (for example a forage harvester or a Rietz Prebreaker). This is used as the feed for a second machine such as a Rietz Vertical Disintegrator fitted with an appropriately sized screen, dependent on the seaweed being used. The product from a Rietz Disintegrator is a slurry of weed and water. The water may be separated using a centrifuge or a rotary drum screen.

2. ACID TREATMENT

In brown seaweeds alginic acid is present mainly as the calcium salt of alginic acid, although magnesium, potassium and sodium salts may also be present. Figure 6 shows that the first major aim of the process is to convert the insoluble calcium and magnesium salts into soluble sodium alginate. If the seaweed is treated with alkali (usually sodium carbonate) then the process necessary for extraction is an ion exchange.

$$Ca(Alg)_2 + 2Na^+ \rightarrow 2NaAlg + Ca^{++}$$

However it has been shown that a more efficient extraction is obtained by first treating the seaweed with dilute mineral acid (Haug, 1964).

$$Ca(Alg)_2 + 2H^+ \rightarrow 2HAlg + Ca^{++}$$

$$HAlg + Na^+ \rightarrow NaAlg + H^+$$

The calcium alginate is converted to alginic acid and this is more readily extracted with alkali than the original calcium alginate; extraction can even be completed at a pH less than 7 (Haug, 1964). At the same time the mineral acid removes all the acid-soluble phenolic compounds. The removal of phenolic compounds is important because (a) they form brown oxidation/polymerisation products with alkali and are largely responsible for a brown discolouration which occurs during alkaline extraction, (b) they cause a loss of viscosity of alginate during alkaline extraction. Pretreatment (i.e., before alkaline extraction) of the seaweed with acid gives a more efficient extraction, a less coloured product and reduced loss of viscosity during extraction, because less of the phenolic compounds are present. Clark and Green (1936) were the first to use this pretreatment.

In practice, the seaweed is stirred with 0.1M sulfuric acid or hydrochloric acid for 30 minutes; temperatures used range from room temperature to about 50°C depending on the seaweed used. Little degradation of alginate occurs with most species of seaweed at temperatures up to 40-50°C. The slurry of seaweed and acid can be separated on a rotary drum screen. The acid-treated weed is usually green and the solid is more free flowing than the untreated material.

3. FORMALDEHYDE TREATMENT

It has been found that acid pretreatment does not remove all phenolic compounds and discoloration still occurs during alkaline extraction. The discoloration can be further reduced by pretreatment with formaldehyde. This process, first used by Le Gloahec (1939), was more thoroughly investigated by Haug (1964) who found that the phenolic compounds and formaldehyde react to give insoluble products, so that no phenolic groups are available for polymerization to dark coloured products during the alkaline extraction.

In practice, the seaweed is stirred with water containing 0.1-0.4% commercial formalin solution, usually at room temperature. Higher temperatures, to about 50°C, can be used but they do not always give a better result. If the particle size of the seaweed has been reduced, as previously described, a reaction time of 15-30 minutes is sufficient. The required concentration of formaldehyde depends on the seaweed being used; some experimentation is necessary to obtain the best conditions for a particular raw material. After treatment, the seaweed is separated using a rotary drum screen and the solids are used in the alkaline extraction.

4. ALKALINE EXTRACTION

In this step, the purpose is to convert the alginate to a soluble form so that it can be removed from the rest of the seaweed. This step can also be used to control the viscosity of the final product. Higher temperatures and longer extraction times lead to breakdown of uronic acid chains and consequent lower viscosities for the sodium alginate. Green (1936) patented a process which used no heating in the alkaline extraction and obtained very high viscosity alginates. The value of producing very high viscosity alginate is debatable; the dried product (usually 10% moisture) is much more prone to breakdown and loss of viscosity, on storage from 6-12 months, than a medium viscosity alginate. Some manufacturers therefore produce medium (and lower) viscosity alginates and for applications requiring very high viscosity, they ensure their product contains sufficient calcium ions to produce the necessary viscosity (see "Properties" section for effect of calcium ions on viscosity). Usually sodium carbonate (soda ash) is used as the alkali because of its low cost; less is required if the seaweed has been given an acid pretreatment.

In practice, seaweed is stirred in a tank with the sodium carbonate solution (about 1.5%) at temperatures from 50-95°C for 1-2 hours. For weed that has undergone size reduction and acid pretreatment, 2 hours at 50°C will generally give good extraction with little degradation of alginate. The time can be reduced by using higher temperatures, usually with some loss of viscosity in the final product. When lower viscosity alginate is desired, the balance of high temperatures versus time can be used to control the viscosity. If whole weed or large pieces are used for the extraction, longer times will be necessary for complete extraction. As extraction proceeds, the extract becomes thicker and may have the consistency of a heavy porridge when complete.

5. SEPARATION OF INSOLUBLE SEAWEED RESIDUE

   A. FLOTATION

The dissolved sodium alginate must now be separated from the alkali-insoluble seaweed residue, which is mainly cellulose. The residue is usually slimy and finely divided. It rapidly clogs filter cloths; uneconomical quantities of precoat material, such as kieselguhr or perlite, must be added to achieve reasonable rates of filtration. Some of the residue can be removed using centrifuges but the clarity of the resulting solution is usually poor. The major portion of the insoluble residue is usually removed by a flotation process, based on that originally described by Le Gloahec (1938). The extract is diluted with 4-6 times its volume of water, to produce a suitable viscosity, about 25-100 cps. Then a small quantity of flocculant is added, air is forced into the liquid, and it is left to stand for several hours. The fine particles of insoluble residue form flocs and are raised to the surface by the rising air bubbles which adhere to the flocs. The residue is scraped from the surface and the clarified liquor, beneath it, is drawn off. The dilution of the original extract must be such as to give a viscosity which allows the particles to rise within an acceptable processing time. This is a very economical and effective method of clarification but the resulting solution is still cloudy. It may require no further clarification if the final product is a technical grade alginate where clarity and colour are not important. However for food grade alginates, a filtration step is usually necessary as well. Because the bulk of the insoluble residue has been removed, it is now economical to use a precoat filter. For very high grade products a second filtration may be used.

In practice, the exact procedure and conditions will vary with the type of seaweed being used since the nature of the insoluble residue will vary. Dilution of the alkali extract is best done by in-line mixing, as is the addition of flocculant. The air can be drawn in via centrifugal-type pumps further down the same line and the diluted, aerated extract is pumped into large holding tanks. The cellulose residue usually has a negative charge so cationic flocculants are used, such as the polyacrylamides available from Allied Colloids under the trade name of Magnafloc. Most suppliers of flocculants have a range of anionic, nonionic and cationic products and provide advice on how to evaluate them for particular applications. The flocculant expedites the removal of very fine particles which would otherwise rise too slowly, being too small for air bubbles to attach themselves; the flocculant brings these fine particles together in larger flocs which air bubbles are more likely to encounter and lift.

In a continuous process, the residue can be continually scraped from the surface as the clarified liquor is removed from the lower part of the tank. In a batch process, many holding tanks are used and the clarified liquor is usually drawn off near the bottom of the tank leaving the residue which is washed out and collected separately. The residue sometimes contains significant amounts of soluble alginate which make it economical to attempt recovery. In this case the residue may be mixed with water again, aerated and let stand to separate.

The process can be carried out without the addition of flocculants but it is not as efficient and the product may be more difficult to filter, requiring either further dilution or heating (up to 50°C); however heating such large volumes can become a significant cost factor.

B. FILTRATION

Any insoluble residue remaining after flotation will be carried through the remaining stages of the process and will appear in the final product. For a final product with good clarity, filtration is required. Because the residue is very fine, filter cloths will rapidly block. The best method is to use a rotary precoat vacuum filter. In this, the rotating drum of the filter is coated with a 2-3 cm layer of precoat material, preferably perlite (an expanded or "puffed" lava, principally aluminium alkali silicate) because it gives a more porous medium than diatomaceous earth (kieselguhr) and so does not block as easily. During filtration, a blade on the rotary filter continually removes the top surface of the precoat, so that a clean filter surface is always available. After 9-10 hours, most of the precoat has been removed by the scraper, filtration is stopped, and a new layer of precoat is deposited. Great care is necessary in selecting the appropriate grade of perlite and the correct cloth to support the precoat medium. Most manufacturers of precoat rotary filters, such as Eimco/Envirotech and Dorr-Oliver, have the facilities to conduct appropriate tests to assist customers or potential customers.

For a very high clarity final product, a second filtration is sometimes used. Usually the quantities involved are smaller and a normal filter press is used. However, a filter aid is still necessary and at this stage a less porous one like diatomaceous earth is suitable. It is usually stirred into the solution to be filtered.

Some manufacturers, particularly those using the alginic acid process, use filters of fine mesh metal or terylene (120 to 200 mesh) to try to clarify the dilute extract obtained from the flotation separation. This is not very successful for most seaweeds although a reasonable result is obtained with some types of Ecklonia. The particles of insoluble residue which remain after flotation are generally too fine to be retained on such sieves so there is little improvement in clarity. This type of filtration may be useful in producing an improved technical grade but, for most seaweeds, the clarity will not be sufficient for food grades.

Some manufacturers have tried to use centrifuges, instead of filtration, to clarify the dilute extract from the flotation separation. Like the use of sieves described above, this may give a better technical grade product but the clarity is insufficient for food grades. One manufacturer, processing Macrocystis, tried to use centrifuges instead of the flocculation-flotation process. The capital outlay was considerable and the result was poor; the solution discharged from the centrifuge had poor clarity and had to be given a flocculation-flotation treatment before it could be economically filtered using precoated filters. Centrifuge performance was improved by extensive dilution of the alkaline extract but the large volumes to be handled and the increased water requirements made it impractical.

6. PRECIPITATION OF CALCIUM ALGINATE

The sodium alginate must be recovered in solid form, once its solution has been separated from the residual seaweed. Evaporation is not practical, the solution is too dilute. The alginate can be precipitated as its calcium salt or as alginic acid, either of which must later be converted to sodium alginate. In this process, calcium alginate is precipitated.

By careful addition of the sodium alginate solution to a calcium chloride solution, calcium alginate can be precipitated in the form of fibres. This fibrous calcium alginate can be readily separated on a metal screen, washed with water and when treated with dilute mineral acid, the Ca^{++} ions are exchanged for H^+ ions, and it yields fibrous alginic acid. This alginic acid can be dewatered using a screw press. Some seaweeds give better fibrous calcium alginate than others; Laminaria gives long fibres which are easier to handle than the short fibres obtained from Ascophyllum.

In practice, it is necessary to add the dilute extract to the calcium chloride solution (about 10%); if the reverse is done, a gel will be obtained instead of fibres. The degree of mixing is important; too little will give a gel-type precipitate while too much may cause excessive breaking up of the fibres, making it difficult to retain them on the metal screen used for separation. The precipitation may be done batchwise in tanks or continuously using an in-line mixer. Operator skill and experience are necessary to obtain consistent results. The fibrous calcium alginate can be separated by running the suspension over a metal screen.

7. BLEACHING

Depending on the seaweed used as raw material, the earlier pretreatments with acid and formalin may ensure a sufficiently pale colour in the final product. However for some food and higher grades, bleaching can be used to improve the colour and odour of the final product, if that is necessary. It is best done at this stage, rather than later, because calcium alginate is more resistant to degradation (loss of viscosity) than alginic acid (Thornley and Walsh, 1934). Usually a sufficient quantity of sodium hypochlorite solution (12%) is added to a suspension of the calcium alginate in water. The quantity of hypochlorite required will vary according to the seaweed used and the effectiveness of the pretreatments with acid and formalin. When a suitably coloured solid is obtained, it is again separated on a metal screen.

8. CONVERSION OF CALCIUM ALGINATE TO ALGINIC ACID

The purpose of this step is to obtain a fibrous alginic acid which can be readily separated and dewatered. This requires an ion exchange ($Ca^{++} \rightarrow H^+$) in the calcium alginate and this is achieved by stirring it in a dilute mineral acid, such as HCl.

In practice, this is done by a three-step countercurrent conversion. Three tanks are used. The calcium alginate is added to the first tank which contains acid previously used in the second tank. After stirring for 30 minutes the solid (now a mixture of alginic acid and calcium alginate) is separated on a screen and the liquid is discarded. The solid is fed to the second tank which contains acid previously used in the third tank. The stirring and separation are repeated and the solid is fed to the third tank which contains unused dilute HCl (0.5M). After stirring and separation, the solid, now alginic acid, is washed once with water. The pH in the three conversion tanks should be adjusted if necessary so that it is always less than pH2.

It can be seen that thorough treatment will produce an alginic acid free of calcium ions. However if calcium is required in the final product (to increase its viscosity) then this can be achieved by varying the conditions of this conversion step and limiting the amount of ion exchange which occurs.

9. DEWATERING THE ALGINIC ACID

The chief advantage of this calcium alginate process is that water can be squeezed from the resulting fibrous alginic acid with relative ease. (This is in contrast to the gel type of alginic acid which results from addition of acid to sodium alginate solution, in the alginic acid process). A screw press, such as the Rietz horizontal continuous S-Press, is often suitable for this squeezing and dewatering. The alginic acid is fed into the rotating graduated-pitch screw and it is compressed by the screw and fixed resistor bars; a screened cone at the end, which rotates at a different speed from the screw, completes the compression and ensures adequate discharge of solids from the press. The dewatered product should contain at least 25% solids if it is to be used in the paste conversion of the next step.

10. CONVERSION OF ALGINIC ACID TO SODIUM ALGINATE

The sodium alginate from the original alkaline extract has now been purified and concentrated in the form of solid alginic acid. This must be converted to solid sodium alginate. Water or an alcohol is used as the 'solvent' with quite different results. In the calcium alginate process, water is usually used. The use of alcohol is described in the alginic acid process.

In practice, the dewatered alginic acid, usually containing greater than 25% solids, is mixed with solid alkali, normally sodium carbonate, in a mixer suitable for blending heavy pastes. The sodium alginate goes into solution, as it forms, in the small amount of water present, giving a heavy paste. However if the original alginic acid had less than 25% solids, the resulting paste may be too fluid. The neutralization can be readily controlled and the product is homogeneous. The reaction can be heated to $50°C$ if necessary. This paste is forced through small holes and the extrusions are chopped into pellets which are dried. They can be dried on trays in a hot-air oven. On a large scale, it is better to use a fluid-bed dryer fitted with a vibrating tilted screen so that the pellets, continuously fed in, vibrate down the screen and out, as the hot air blows up through the screen. The dried pellets (about 10% moisture) can be milled to an appropriate particle size, usually about 60 mesh (250 microns).

ALGINIC ACID PROCESS

Stages 1. to 5. are identical to the same stages in the calcium alginate process.

6. BLEACHING

Treatment with sodium hypochlorite (12% solution) is best done under alkaline conditions, there is less degradation of the alginate chains, so it is sometimes added to the clarified/filtered alkaline extract (Henkel, 1964; Okazaki, 1971). However the volumes of solution are very large at this stage, because of the dilutions which were necessary in the separation of the insoluble residue in stage 5, so some manufacturers find it more economical to add the hypochlorite in the final step of the process, the conversion of alginic acid to sodium alginate, which is frequently done using sodium hydroxide and an alcohol solvent. The quantity of hypochlorite used is kept to a minimum for economic reasons and will depend on the colour of the original seaweed used for the process, the effectiveness of the acid and formalin pretreatments, and the amount of insoluble, coloured solids which remain after stage 5.

7. PRECIPITATION OF ALGINIC ACID

The clarified sodium alginate extract is treated with dilute mineral acid, usually HCl or H_2SO_4, at room temperature and a gelatinous precipitate of alginic acid forms which cannot be filtered, it simply blocks any filter medium. It can be removed from the liquid by flotation. Usually a slight excess of sodium carbonate is left in the alkaline extraction filtrate so that the acid addition releases carbon dioxide gas; this becomes incorporated in the alginic acid gel which forms simultaneously. The gas lifts the alginic acid to the surface. The alginic acid gel becomes firmer if a layer several inches thick is allowed to build up on the surface. Once such a layer has formed, the alginic acid is scraped off. For successful separation, this alginic acid precipitate must be treated gently (a) so that it does not break up into fine pieces and (b) so that the carbon dioxide is not lost from its structure. Losses of alginic acid in the drain liquors can be severe if the flotation is not carried out carefully. The alginic acid precipitate must be left in contact with the mineral acid for sufficient time to allow it to react with any sodium alginate solution occluded in the precipitate formed.

In practice the mineral acid (such as 5% sulphuric acid) and alginate solution may be gently mixed in-line, with a pH controller metering the acid addition to give a final pH of 1.5-2.0; the mixture is left to stand for about 60 minutes in a rubber-lined tank, allowing completion of reaction and flotation of the alginic acid. Le Gloahec and Herter (1938) describe a procedure, variations of which are still used; the principle is to mix the acid (usually a small volume) and the alginate solution (usually a large volume) as they flow over an inclined baffle so that good, gentle mixing is achieved with a minimum of occlusion of the alkaline sodium alginate solution. Sometimes the mixture is allowed to flow along a sloping, long, baffled channel, to allow further mixing and reaction, before running into the tank used for flotation. After the alginic acid is removed from the flotation tank, the remaining solution is run to waste but its content of alginic acid must be periodically monitored to ensure that losses at this stage are minimal.

8. DEWATERING ALGINIC ACID

This is undoubtedly the most difficult stage of the alginic acid process. The alginic acid gel obtained from flotation contains only 1-2% solids; its conversion to sodium alginate would give a viscous solution whereas a solid, or a paste which could be dried to a solid, is required. Some of the water can be removed from the gel by (a) pressing or squeezing, (b) centrifuging, (c) mixing with an alcohol.

(a) The consistency of the gel scraped from the flotation tank is too soft to allow the use of a screw press. The consistency can also vary from hour to hour and day to day; it may be quite firm, or it may be soft or sloppy. Some manufacturers, particularly those using a squeezing process, hold it for 1-2 hours in large containers of coarse filter cloth; this allows some of the liquid to drain away. The material can then be shovelled into smaller nylon/terylene filter cloth bags, which are closed and stacked between the plates of a hydraulic press. Pressure is applied and liquid is squeezed from the gel. The alginic acid, now about 20% solids, is removed from the bags and may be used for the next stage, conversion to sodium alginate, if alcohol is to be used as the solvent; if water is to be the solvent, the squeezing process is repeated to give a product with about 25-30% solids (see stage 10 of the calcium alginate process). This process is labour intensive and only economical in some countries. Another method of squeezing is to use a sequence of pairs of rollers, with a steadily decreasing clearance between each pair. Two continuous belts of filter cloth feed the gel between the rollers where it is squeezed and the liquid escapes through the belt. This system requires careful control of the consistency of the feed material; if the gel is too soft it can be squeezed out the sides of the belts.

(b) Some manufacturers take the gel straight from the top of the flotation tank and place it in basket-type centrifuges with filter cloth liners on the inside. Centrifuging can increase the solids content to 7-8%; this is suitable if alcohol is to be used in the conversion process but this material is also sufficiently firm for it to be further dewatered using a screw press.

(c) Dewatering the alginic acid from flotation by mixing with an alcohol (usually methanol or ethanol) is suitable as a laboratory procedure but is uneconomic on a large scale because of the costs (i) of recovery of the alcohol and (ii) of the alcohol lost, since recovery is rarely complete. Sometimes, however, alcohol is used to further dewater the alginic acid obtained after squeezing.

10. CONVERSION OF ALGINIC ACID TO SODIUM ALGINATE

Alginic acid with 25% or more solids can be converted, using water as the solvent, by the paste method described in the calcium alginate process. Most manufacturers using the alginic acid process tend to use alcohol as the solvent for conversion because the water content of the alginic acid is high.

The alginic acid is suspended in alcohol (methanol or ethanol) at room temperature and a strong sodium hydroxide solution (40%) is added. The exchange of H^+ and Na^+ is slow since neither the alginic acid nor the sodium alginate is appreciably soluble. It is not easy to obtain a homogeneous neutralization because this depends on how well the alkali can penetrate the particles of alginic acid. The reaction is followed by taking samples of the solid and measuring the pH of its aqueous solution. The pH of the alcohol solvent is not a measure of the degree of the conversion. When the sodium alginate has pH 6, the suspension can be filtered or centrifuged.

Usually the sodium alginate produced is in the form of fine hair-like fibres, because that was the form of the original alginic acid precipitate. The thickness of these fibres imposes limitations on the maximum particle size which can be obtained after grinding. This is one of the disadvantages of the alcohol conversion method. Fibres and fine powders of sodium alginate are difficult to dissolve because neither disperse easily in water. Ease of solubility is a very important factor for customer acceptance. A more granular or coarse powder, which dissolves more readily, can be obtained from grinding the pellets obtained from the paste method of conversion.

The solid sodium alginate from the filter is sometimes squeezed in a screw press, especially if the alginic acid was only 7-8% solids. This removes most of the residual alcohol-water before it is sent to the dryer. It is dried to about 10% moisture and milled to appropriate particle sizes, according to the intended application. The drying should be in a system equipped for the recovery of the alcohol vapour. This recovered alcohol can be added to the alcoholic filtrate which is distilled and recycled. Alcohol is a relatively expensive raw material and the economical production of alginate using alcohol conversion is very dependent on having a high percentage recovery of alcohol.

An alternate method of alcohol conversion, suitable if the alginic acid has about 20% solids or more, is to place the alginic acid in a paste-type mixer, add strong sodium hydroxide solution and then just enough alcohol to allow the mixing of the wet fibrous solid. When the reaction is complete (10-15 minutes, pH 5.5-6.0) the wet fibrous solid is drained, squeezed in a hydraulic or screw press and then dried. Much smaller quantities of alcohol are used in this method but its recovery is still an important economic factor.

GENERAL

1. WATER

Alginate factories must be established near an adequate water supply since the requirements are high and can range from 1 000-1 500 m^3 per tonne of final product. The water should be clear and free from any colloidal clay and suspended matter. The presence of calcium and magnesium ions can cause formation of the corresponding alginate salts which will cause difficulties in some stages of the process. Interference from these ions can be prevented by pretreatment of the water with ion-exchange resins. Bacterial levels should be low, particularly for the production of food and pharmaceutical grades.

2. BACTERIAL CONTAMINATION OF EQUIPMENT

Many brown seaweeds have a natural flora of bacteria which possess enzymes, alginate lysases, capable of breaking down the alginate molecule. These bacteria inevitably enter the processing equipment with the seaweed so precautions must be taken to ensure they do not proliferate, otherwise there will be serious losses in the viscosity of the alginate produced. Seaweed and process residues must not be allowed to accumulate, especially in concrete tanks which have a porous surface and are difficult to clean.

DERIVATIVES

SALTS

Sodium alginate is the main form of alginate in use. Smaller quantities of alginic acid and the ammonium, calcium, potassium and triethanolamine salts are also produced. Calcium alginate and alginic acid are made during the calcium alginate process for making sodium alginate; each can be removed at the appropriate stage, and after thorough washing, can be dried and milled. The other salts are made by neutralization of moist alginic acid with the appropriate base; sufficient water or alcohol can be added to keep the material at a workable consistency and it is processed as described for the paste-conversion method in the calcium alginate process. However the triethanolamine salt is hygroscopic and is best dried in thin layers and then milled.

PROPYLENE GLYCOL ALGINATE

Propylene glycol alginate was first prepared by Steiner (1947) but a better method was published in a more informative patent by Steiner and McNeely (1950). It is made by the reaction of propylene oxide with alginic acid and the carboxylic acid groups on the uronic acid chains are esterified. The original patent (Steiner, 1947) gave a product with a pH of about 3 but it had poor viscosity stability, both in the solid state and in solution. However when a partially neutralized alginic acid was used, the reaction was accelerated and yielded a more stable product with a pH 3.8-4.6; under these conditions less hydrolysis of both the alginic acid and propylene oxide occurred (Steiner and McNeely, 1950). This patent gives a useful description of the process.

Partially neutralized alginic acid can be made by reacting 5-20% of the carboxylic acid groups with an alkali and the product has a pH 3.5-5.5. Ammonia is used for the neutralization because the excess is easily removed, but some methods for using sodium carbonate or sodium phosphate are also described. The alginic acid and alkaline reagent are mixed using repeated passes through a hammer mill and by passing a current of warm, dry air through the mill, the moisture content can be adjusted to 45-55%. This moisture content gave the best reaction rate for esterification with the least hydrolysis of the propylene oxide. The fibrous alginic acid and gaseous propylene oxide, in a mole ratio of about 1:3, are mixed in a pressure vessel at 45-60°C. The reaction takes 8 hours at 50°C and gives a product with about 80% of the carboxyl groups esterified and pH 3.9.

An improvement in the above process was reported by Pettitt and Noto (1973). They were able to reduce the reaction time to 2-3 hours, mainly by removing any inert gas such as air from the reaction vessel. The air is either removed by vacuum or purged by a gas flow of propylene oxide. The alginic acid is partially neutralized to 8-22%, as described by Steiner and McNeely (1950), but its solids content is more carefully controlled at 65-78%; if the solids fall below 65%, the increased moisture leads to excessive formation of propylene glycol and if it rises above 78% the reaction rate becomes very slow. The reaction can be run from 60-100°C with reaction times of 2-3 hours, to give a product (approximately 80% mole ester, pH 3.8-4.6) which has good viscosity stability at room temperature for several months.

More recently Noto and Pettitt (1980) described a process using liquid propylene oxide mixed with partially neutralized alginic acid in a pressure vessel. Good esterification is achieved even with a very low neutralization (0.4%) of the alginic acid and the latter can contain as little as 20% solids, although most of the examples cited contain 34% solids. This means much less drying of the alginic acid is necessary before use in the reaction. The reactions were generally at 75-85°C and required only 2 hours.

PROPERTIES

The following discussion centres on those properties of alginates which are particularly relevant to their uses. Information about other properties, and sometimes more detailed information about the properties discussed here, can be found in Kelco (1976), McDowell (1977), Cottrell and Kovacs (1980) and King (1983).

The commercial products of most interest are sodium alginate, propylene glycol alginate and alginic acid; smaller quantities are used of the potassium, ammonium, calcium and triethanolamine salts as well as mixed salts of sodium and calcium.

STABILITY-SOLID ALGINATES

The degree of polymerization (DP) of an alginate is a measure of the average molecular weight of the molecules and is the number of uronic acid units per average chain. DP and molecular weight relate directly to the viscosity of alginate solutions; loss of viscosity on storage is a measure of the extent of depolymerisation of the alginate.

Sodium alginate is produced in various grades, usually described as low, medium and high viscosity alginates (this refers to the viscosity of its 1% aqueous solution).

Alginates with a high DP are less stable than those with a low DP. Low viscosity sodium alginates (up to about 50 mPa.s) have been stored at 10-20°C with no observable change in 3 years. Medium viscosity sodium alginates (up to about 400 mPa.s) show a 10% loss at 25°C and 45% loss at 33°C after one year, and higher viscosity alginates are less stable.

Propylene glycol alginates show about 40% loss in viscosity after a year at 25°C and also become less soluble. Ammonium alginate is generally less stable than any of the above. Alginic acid is the least stable of the products and any long chain material degrades to shorter chains within a few months at ambient temperatures. However short chain material is stable and alginic acid with a DP of about 40 units of uronic acid per chain will show very little change over a year at 20°C. However the main use of alginic acid, as a disintegrant in pharmaceutical tablets, depends on its ability to swell when wetted and this is not affected by changes in DP.

The commercial alginates should therefore be stored in a cool place, 25°C or lower, since elevated temperatures can cause significant depolymerization which affects the commercially useful properties such as viscosity and gel strength. They usually contain 10-13% moisture and the rate of depolymerization increases as the proportion of moisture is increased, so the storage area should be dry.

STABILITY - ALGINATE SOLUTIONS

The monovalent cation salts [Na^+, K^+, NH_4^+, $(CH_2OH)_3NH^+$] of alginic acid, and its propylene glycol ester, dissolve in water but alginic acid and the calcium salt do not. Neutral solutions of low to medium viscosity alginates can be kept at $25°C$ for several years, without appreciable viscosity loss, as long as a suitable microbial preservative is added. Solutions of highly polymerized alginates will lose viscosity at room temperature within a year and to achieve high, stable viscosities it is better to add calcium ions to a solution of an alginate with a moderate DP. All solutions of alginate will depolymerize more rapidly as the temperature is raised. Alginates are most stable in the range of ph 5-9 (McDowell, 1977). Small amounts of calcium greatly increase the stability of sodium alginate solutions. Propylene glycol alginate solutions are stable at room temperature from pH 3-4; below pH 2 and above pH 6 they will lose viscosity quickly even at room temperature (McNeely and Pettitt, 1973).

Since alginate solutions contain a polysaccharide anion, they cannot be mixed with cations which will combine with this anion to give an insoluble product. Alginate solutions are incompatible with most divalent and trivalent cations, with quaternary ammonium salts such as those used as bactericides, with acids strong enough to cause precipitation of alginic acid, with strong alkalis which lead to a gradual breakdown of the polysaccharide chains. The compatibility of a more specific list of substances, which are likely to be used with alginates, is discussed by Kelco (1976).

SOLUBILITY

A. PHYSICAL FACTORS

When powders of soluble alginates are wetted with water, the hydration of particles results in each having a tacky surface. Unless some precautions are taken, the particles will rapidly stick together resulting in clumps which are very slow to completely hydrate and dissolve. Particle size and type affect solubility behaviour.

Coarse particles are usually preferred because they are easier to disperse and keep separate, even though they are slower to hydrate and dissolve. Fine particles will dissolve more rapidly but there is more risk of them clumping together; this risk is less if the alginate is diluted with another powder such as a sugar. Alginates made using an alcohol conversion step (described in "Extraction Processes" section) often have fibrous particles; these usually hydrate more rapidly than granular particles (resulting from paste conversion) but tend to wind around each other and are more difficult to disperse.

The quantity of the soluble alginates which will dissolve in water is limited by the physical nature of the solutions rather than actual solubility. As the concentration of alginate increases the solution passes through stages of a viscous liquid to a thick paste; at this point it becomes very difficult to disperse further alginate successfully.

B. CHEMICAL FACTORS

It is more difficult to dissolve alginate in water if the water contains compounds which compete with the alginate for the water necessary for its hydration. The presence of sugars, starches or proteins in the water will reduce the rate of hydration and longer mixing times will be necessary. Salts of monovalent cations (such as NaCl) have a similar effect at levels above about 0.5%. All of these substances are best added after the alginate has been hydrated and dissolved. The presence of small quantities of many polyvalent cations inhibits the hydration of alginates and larger quantities cause precipitation. Sodium alginate is difficult to dissolve in hard water and milk because both contain calcium ions; these ions must first be sequestered with a complexing reagent such as sodium hexametaphosphate or ethylenediamine tetraacetic acid (EDTA). Propylene glycol alginate (preferably 80-85% esterified) is less affected by calcium ions and can be used in milk. Acidic conditions also inhibit hydration and when the pH is less than 4.0 it is better to use propylene glycol alginate which remains soluble down to about pH 2.

Alginates are insoluble in water-miscible solvents such as alcohols and ketones. Aqueous solutions (1%) of most alginates will tolerate the addition of 10-20% of these solvents; propylene glycol alginate tolerates 20-40% while up to 65% ethanol can be added to triethanolamine alginate without causing precipitation. The presence of such solvents in water, before dissolving the alginate, will hinder hydration.

C. DISSOLVING ALGINATES

Most alginate manufacturers provide detailed information on how best to dissolve alginates. The advice of Kelco (1976) is typical; they suggest methods based on (a) high-shear mixing, (b) dry-mix dispersion, (c) liquid-mix dispersion.

In high-shear mixing, the principle is to prevent the clumping together of the particles, which become tacky as soon as the surface is hydrated. Powdered alginate is slowly poured into the upper part of a vortex created in the water by high speed stirrer; the stirrer blades must remain submerged to avoid too much aeration. If some clumps do form, the shear should be sufficient to break them up. For large scale mixing, Kelco sell a funnel attached to a mixing aspirator; a fast flow of water through the aspirator sucks in alginate powder from the funnel, mixes and wets it, and then discharges it into a well agitated tank of water.

Dry-mix dispersion can be used when a formulation requires both alginate and other dry ingredients such as sugars, starches, etc. The dry powders are mixed thoroughly so that the alginate particles are diluted and separated by the other ingredients. This mixture is slowly added to well stirred water, preferably with a vortex as before, and the other ingredients, often in a ratio of 5:1 to 10:1, help to keep the alginate particles apart as they are wetted.

An even more efficient method of diluting the alginate particles is to use liquid-mix dispersion in which they are wetted with a non-solvent. This can be either a water-miscible non-aqueous liquid (such as ethanol or glycerol) or a water-immiscible liquid (such as a vegetable oil). Enough liquid is needed to give a pourable slurry and this is poured into the water, well agitated as before. The particles are dispersed and the rate of hydration, and solution, will depend on the time taken for the non-solvent liquid to diffuse from the surface of the particles.

D. PRESERVATION OF SOLUTIONS

Microorganisms will grow in solutions of commercial alginates because they usually contain sufficient nitrogenous compounds and salts. Bacterial or mould growth may cause depolymerization and loss of viscosity of the alginate as well as contamination and spoiling of any product in which the alginate is used. Food and cosmetic products are protected by their traditional preservatives such as sorbic acid, potassium sorbate, benzoic acid, sodium benzoate and the methyl or ethyl ester of p-hydroxybenzoic acid. For other uses less expensive, and sometimes more effective, preservatives are available such as formaldehyde and sodium pentachlorophenate and other phenol derivatives. Compounds of copper and zinc, and quaternary ammonium salts, should not be used because they will react with the alginate.

VISCOSITY

Many of the uses of alginates depend on their thickening effect, their ability to increase the viscosity of aqueous systems using relatively low concentrations. At the concentrations used in most applications, the viscosity behaviour of alginate solutions is pseudoplastic, the solution flows more readily the more it is stirred or pumped (the viscosity decreases as the rate of shear increases). This effect is reversible except at very high rates of shear (Glicksman, 1969). It is most marked with high molecular weight alginates, with sodium alginate solutions which contain calcium ions, and with propylene glycol alginate above 1% concentration; some of these solutions can also be thixotropic, that is they show a time-dependent thinning at constant shear rate and their recovery to the initial viscosity is time dependent. Reproducible viscosity measurements are made using a rotational type of viscometer, such as the Brookfield Synchro-Lectric.

Several factors influence the viscosity of alginate solutions.

A. MOLECULAR WEIGHT

The higher the molecular weight of a soluble alginate, the greater the viscosity of its solution. Manufacturers can control the molecular weight (degree of polymerization, DP) by varying the severity of the extraction conditions and they offer products ranging from 10-1 000 mPa.s (1% solution) with a DP range of 100-1 000 units. Sodium alginate of viscosity 200-400 mPa.s, "medium viscosity", probably finds the widest application.

B. CONCENTRATION

There is no simple relationship between concentration and viscosity for alginate solutions but McDowell (1960,1977) found a useful empirical equation which applied to a wide variety of alginates over a range of at least a hundredfold change in viscosity:

$$\log_{10} \text{viscosity} = a\sqrt{(\text{concentration})} - b$$

where a is a constant related to the DP of the alginate, b is a constant for a particular type of alginate. Graphs or tables of viscosity versus concentration are available from manufacturers for their particular products (Kelco, 1976; McDowell, 1977; Protan, 1986) and some typical figures are shown in Table 3.

C. TEMPERATURE

Viscosity decreases as temperature increases, at a rate of about 2.5% per degree Celsius (Figure 7). Viscosity usually returns to a little less than the original value on cooling. However if alginate solutions are kept above 50°C for several hours, depolymerization may occur giving a permanent loss of viscosity.

Alginate solutions can be frozen and thawed without change of viscosity, as long as they are free of calcium (less than 0.5%); if calcium is present the viscosity will increase and a gel may even form and these changes will not reverse.

D. pH

The viscosity of alginate solutions is unaffected over the range of pH 5-11. Below pH 5, the free -COO⁻ ions in the chain start to become protonated, to -COOH, so the electrostatic repulsion between chains is reduced, they are able to come closer and form hydrogen bonds, producing higher viscosities (King, 1983). When the pH is further reduced, a gel will form, usually between pH 3-4; however if the alginate contains residual calcium this gelation may occur about pH 5. If the pH is reduced quickly from pH 6 to pH 2, a gelatinous

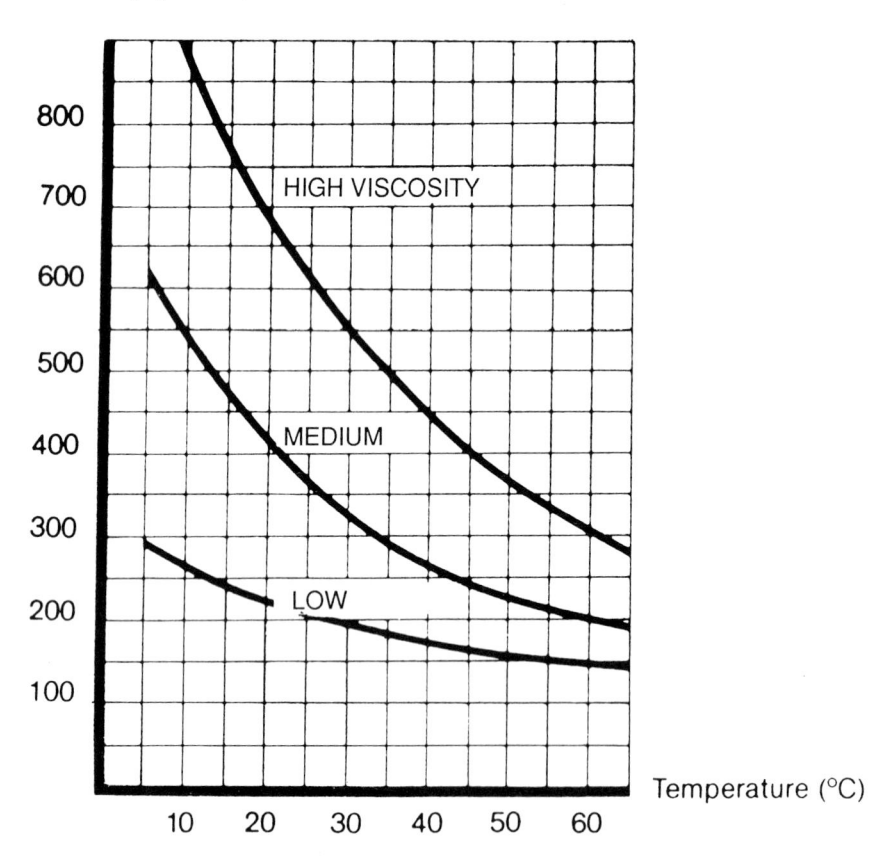

Figure 7 Viscosity in 1% solutions at different temperatures. (Source: Protan, 1986a)

precipitate of alginic acid will form. Above pH 11, slow depolymerization occurs on storage of alginate solutions, giving a fall in viscosity.

Propylene glycol alginate has fewer $-COO^-$ ions and is less affected by increasing acidity. Its solutions remain unchanged to about pH 3; below this value precipitation and gel formation occur. Above pH 6.7, hydrolysis of the ester groups occurs slowly with consequent loss of viscosity.

E. CALCIUM IONS

The presence of low concentrations of calcium ions in an alginate solution will increase its viscosity and larger amounts will cause the formation of a gel. The addition of Ca^{++} is therefore a way of increasing the viscosity of a solution without having to increase either the amount of alginate dissolved or the molecular weight of the alginate being used. It also allows the flow properties of solutions to be adjusted (reduced) by adding sequestering agents such as calgon and EDTA. The disadvantage is that alginate solutions with calcium ions show a greater loss of viscosity with stirring (are more shear sensitive) than alginates with no calcium. As the concentration of calcium ions and viscosity increase, the solutions change from pseudoplastic to thixotropic, that is they take some time to recover their original viscosity after being stirred. The way in which calcium reacts with alginate is discussed in the later section on "Gels".

Most commercial alginates made by the calcium alginate process contain residual quantities of calcium; for example the usual food grades of sodium alginate from Kelco contain 1.2% calcium and in special low-calcium grades this is reduced to 0.2% (Kelco, 1976). 1.2% calcium represents 17% substitution of calcium for sodium in the sodium alginate and this is sufficient to increase the viscosity; thickened, flowable solutions result from 7-20% substitution by calcium ion while gels form with about 30%. There is a region of calcium addition, just before gel formation, where very thick solutions result which are thixotropic. The range of calcium concentration over which this occurs is much greater for alginates with a high content of mannuronic acid (high M/G ratio). Alginates with a high guluronic acid content show a more abrupt transition from solution to gel. The effect of calcium on the viscosity of a calcium-containing alginate can be estimated by measuring the viscosity of a solution before and after the addition of a sequestering agent, such as sodium hexametaphosphate, which removes the calcium ions in a complexed phosphate ion.

Alginates made by the alginic acid process contain negligible amounts of calcium so that if an increase in viscosity is required at a fixed alginate concentration, a small amount of a sparingly soluble calcium salt such as calcium sulfate or calcium citrate may be added. The effect of calcium on the viscosity of an alginate is difficult to predict and is usually found by experimentation. It will depend on the uronic acid composition and degree of polymerization of the alginate; alginates with higher molecular weights and/or higher M/G ratios give greater viscosity changes (McDowell, 1960). The way in which the solution is made also affects the final viscosity; for example a 1% solution prepared by dilution of a 3% solution will differ in viscosity from a 1% solution prepared directly; the type and duration of the stirring used in preparation will also affect the result.

Propylene glycol alginate with 85% of the carboxylic acid groups esterified is hardly affected by the presence of calcium ions. At the other extreme, 60% or less esterification gives an ester which behaves similarly to sodium alginate except that thixotropic effects are much more evident.

GELS

The polysaccharides derived from seaweeds - alginates, agars, carrageenans and furcelleran - can all be induced to form gels under certain conditions. A better understanding of the structure of these gels has developed in recent years and useful reviews have been written by Rees (1972) and Morris (1985).

Solutions of alginate will react with many di- and trivalent cations to form gels; the gels will form at room temperature, or any temperature up to $100^{\circ}C$, and they do not melt when heated. They find applications in various industries, particularly when calcium is used as the divalent ion. Alginate solutions will also form gels if they are carefully acidified; these gels are generally softer than calcium gels and, unlike calcium gels, give the feel of melting in the mouth so they find many applications in the food industry.

CALCIUM GELS

Those who are seriously interested in formulating calcium gels should refer to the work by King (1983, pp. 141-173) which is the most thorough discussion available of the variables which should be considered, the practical systems commonly used and some examples of actual applications.

Calcium has found greatest popularity as the divalent ion for gel formation because its salts are cheap, readily available and non-toxic. If a calcium chloride solution is stirred into an alginate solution, a precipitate of calcium alginate results; it may be stringy or gelatinous. To obtain a smooth gel, the calcium must be released slowly into the alginate solution. This is done by using a calcium salt with a low solubility (such as calcium citrate) which slowly releases calcium ions. An alternate method is to use a calcium salt which is practically insoluble in neutral solution but dissolves as the pH falls (such as dicalcium phosphate); when an acid of low solubility (such as adipic acid) is added, it gradually lowers the pH, calcium ions are released and a gel forms. The time needed for a gel to form can be controlled by the solubilities of the calcium salt and acid, their particle size and the operating temperature. Retarding agents can also be used, such as sequestrants which complex the calcium ions and make them unavailable until all the sequestrant has reacted; when a dry powder product contains both alginate and a calcium salt, the addition of sufficient sequestrant will delay the availability of calcium ions until the alginate is dispersed and hydrated. Details of calcium salts, acids and sequestrants which are used have been discussed by Littlecott (1982) and more detail by King (1983).

The gel strength depends on the source (algal species) of the alginate, the concentration of alginate, its degree of polymerization and the calcium concentration. Alginates from different seaweeds can have differing ratios of mannuronic acid to guluronic acid in their structures and different proportions of M, G and MG blocks (see "Structure of alginic acid" section). This ratio, and the way in which the acids are distributed in the alginate chains, have a marked effect on gel formation and gel strength. Alginates with a high proportion of G blocks form rigid gels; they form fairly suddenly as calcium ion concentration is steadily increased. The opposite holds for alginates with mainly M blocks; they form gradually and are softer and more elastic. This behaviour is related to the molecular structure of the gels.

The early hypotheses for gel formation was that calcium ions displaced hydrogen ions on the carboxylic acid groups of adjacent chains and formed simple ionic bridges between the chains. Rees (1969) argued why that was unlikely and later he put forward the "egg-box model" (Grant et al., 1973), now generally accepted. This requires the cooperative mechanism of binding, of two or more chains, shown in Figure 8. The buckled chain of guluronic acid units is shown as a two-dimensional analogue of a corrugated egg-box with interstices in which the calcium ions may pack and be coordinated. "The analogy is that the strength and selectivity of cooperative binding is determined by the comfort with which 'eggs' of the particular size may pack in the 'box', and with which the layers of the box pack with each other around the eggs" (Grant et al., 1973). The model can be extended to be three-dimensional. While calcium helps to hold the molecules together, their polymeric nature and their aggregation bind the calcium more firmly; this has been termed "cooperative binding". The structure of the guluronic acid chains (Figure 4) gives distances between carboxyl and hydroxyl groups which allow a high degree of coordination of the calcium. The strontium ion is larger and even more firmly bound; magnesium ions are smaller and are not held, so magnesium alginate does not form gels (McDowell, 1977).

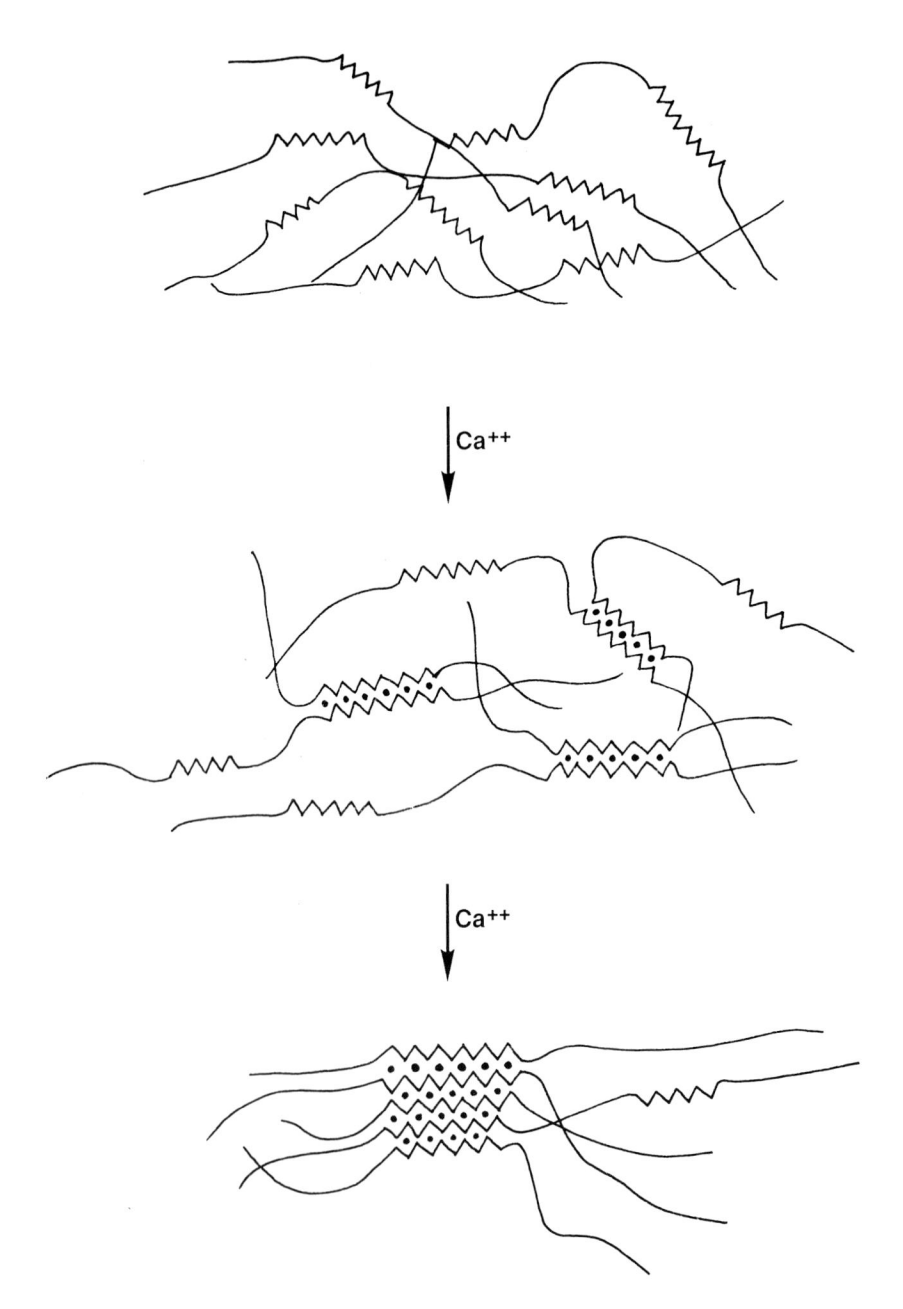

Figure 8 Gel formation via G blocks: egg box model

Propylene glycol alginate, with a low degree of esterification (below 60%) and a high degree of polymerization, can form soft gels with calcium salts. As the degree of esterification is raised to about 85%, it is hardly affected by calcium (McDowell, 1977). It is also more tolerant of calcium at lower pH values (Steiner and McNeely, 1950).

ACID GELS

The structure of these gels has not been studied as comprehensively as calcium gels, probably because they are more limited in their application. A steadily increasing number of carboxyl ions on the alginate chains become protonated as the pH falls, reducing the electrical repulsion between chains. The chains can then move closer together which allows hydrogen bonding to be more effective. At first this produces a higher viscosity and eventually, at pH 3.5-4.0, a gel forms. Small amounts of calcium (less than 0.01%) must be present; the reason is not known.

King (1983) has outlined the useful characteristics of acid gels. For equivalent alginate concentrations, acid gels have about half the strength of calcium gels and they do not show any syneresis. This softness, when combined with their feeling of melting in the mouth, means they are useful in some food applications where they can imitate the effect of gelatin; calcium gels, even soft ones, still feel lumpy in the mouth. They can be made so that they can be stirred or pumped and then reset to a gel, a valuable property when manufacturing some processed foods. However acid gels are not stable when heated and become softer with time even at room temperature, as the alginic acid depolymerizes; they are stable for about a year if refrigerated ($5^{o}C$). Propylene glycol alginate is not suitable for making acid gels.

FILM FORMATION

Alginates can be made into two types of film which have different properties: water-soluble films (usually from sodium alginate) and oil-soluble films (usually from calcium alginate).

Water-soluble films can be made by evaporation of a solution of alginate or by extrusion of an alginate solution into a non-solvent which mixes with water, such as acetone or ethanol. These films are impervious to grease, fats and waxes but allow water vapour to pass through. They are brittle when dry but can be plasticized with glycerol, sorbitol or urea. They have good non-stick properties and are useful as mould release agents, for example in the manufacture of fibreglass plastics. Where a high solids film is needed, a very low viscosity alginate can be used. Self-supporting films need greater strength and require the use of higher viscosity alginates with a greater degree of polymerization. Triethanolamine alginate is used to form soft flexible films.

Water-insoluble films can be made by treating a water-soluble film with a di- or trivalent cation (Ca^{++} is the most frequently used one) or with acid. They can also be made by extrusion of a solution of a soluble alginate into a bath of a calcium salt. Some alginates, such as zinc alginate, are soluble in excess ammonia solution; if the NH_3 is evaporated from a film of such a solution, an insoluble film of zinc alginate remains. These films of insoluble alginate are not water-repellent and will swell on prolonged exposure to water.

FILAMENT FORMATION

If a solution of sodium alginate is forced through fine holes into a solution of a calcium salt, filaments of calcium alginate will be formed. Much research went into the development of alginate yarns (Steiner and McNeely, 1954; Maass, 1959) but they are not resistant to alkaline soaps. This, plus the relative cost of alginate and the development of many synthetic fibres, led to a loss of interest until quite recently when a new commercial product appeared. Made in the United Kingdom, it is a bandage-type material which is used as a dressing on wounds. When it comes into contact with sodium salts in the body fluids, some of the calcium ions are exchanged for sodium ions so that a thin soft gel forms at the interface of the dressing and the wound and the dressing never sticks to the wound.

GENERAL COLLOIDAL PROPERTIES

General colloidal properties is the term used to explain why alginate is successful in some applications where the reasons are not fully understood and where the alginate has been chosen on an empirical basis. These applications have been discussed by McDowell (1960) and Leigh (1979). They include the use of: sodium alginate, as a flocculant, as a suspending agent, and as a stabilizer in ice cream; propylene glycol alginate, in acidic frozen products such as ice sherberts, in fruit squash containing fruit solids, and in stabilizing beer foam.

SAFETY IN FOODS

The Food Chemical Codex gives specifications for alginic acid, its propylene glycol ester and its ammonium, calcium, potassium and sodium salts. These four salts have been granted GRAS status (generally recognized as safe) in the USA and propylene glycol alginate has been approved as a food additive for use as an emulsifier, stabilizer or thickener. The joint Expert Committee of Food Additives of the Food and Agriculture Organization of UN/World Health Organization has also issued specifications for alginates and recommended an Acceptable Daily Intake, for alginic acid salts of 50 mg per kg body weight per day, for propylene glycol alginate of 25 mg/kg/day. King (1983) has listed 39 countries which permitted alginate salts as at January 1982; three of them had not approved the propylene glycol ester. Food additive laws differ from country to country, even from state to state within a country, and are constantly being revised. Therefore users must acquaint themselves with the latest information in their relevant countries and cannot rely on the information given here and in the references cited.

USES

Earlier reviews on the uses of alginates include those by Steiner and McNeely (1954), Maass (1959), McDowell (1960), Glicksman (1969a), McNeely and Pettitt (1973), Cottrell and Kovacs (1980), and King (1983).

Not all reviews cover all uses; for example while most give lengthy treatments to food uses, few say much about textile printing and paper applications. Reviews which are more specific to a particular use are listed in the following subsections. The main uses of alginates are shown in Table 4.

TEXTILE PRINTING

In textile printing, alginates are used as thickeners for the paste containing the dye. These pastes may be applied to the fabric by either screen or roller printing equipment. An excellent review by Hilton (1969) discusses the role of the thickener in the printing of fabrics and the advantages/disadvantages of sodium alginate in different printing processes. Alginates became important thickeners with the advent of reactive dyes which combine chemically with cellulose at its hydroxyl groups. Many of the standard thickeners, such as starch, also react with these dyes and this leads to lower colour yields and sometimes insoluble products which are not easily washed out and which can result in a fabric with poor handle. Alginates react minimally with reactive dyes, they wash out of the finished textile readily and are the best thickeners for these dyes. They are also used with other types of dyes.

The viscosity of the paste can be varied according to the application and the equipment. Thick pastes with short flow characteristics are useful when the extent of penetration into the fabric must be limited but thinner pastes with long flow are required for fine-patterned prints. For alginates containing small quantities of calcium, viscosity can be controlled by adding sequestering agents such as polyphosphates. However these pastes are more likely to lose viscosity as shear rate increases and a paste which is less shear sensitive can be made using a high concentration of a lower viscosity alginate. This latter kind of paste is especially useful for printing disperse dyes on synthetic fibres. Most alginate manufacturers can supply basic recipes for the different types of dyes and printing processes (for example, Protan, 1985) which are a useful starting point; the quantities of alginate can vary from 1.5% of high viscosity alginate to 5% of low viscosity alginate.

Table 3

Variation of viscosity (mPa.s) with concentration for sodium alginate solutions at 20°C

Type of alginate	Concentration				
	1%	1.5%	2%	3%	4%
Very low viscosity	10	20	45	130	350
Low viscosity	20	60	180	650	2200
Medium viscosity	350	1800	6000	not measurable	
High viscosity	800	4000	9000	not measurable	

Table 4

Principal uses of alginate

End-uses	Percentage of the quantity of total demand
Textile printing	50
Food	30
Paper	6
Welding rods	5
Pharmaceuticals	5
Others	4

Source: ITC (1981).

Alginates are normally incompatible with cationic dyes. However Racciato (1979) has reported that premixing the cationic dye with selected surfactants before addition to the thickener will allow the use of many cationic dyes. He claims that compatibility with almost every cationic dye can be obtained if either xanthan gum or algin is used. In the printing of cotton cloths using reactive dyes, Prelini (1982) suggests ways of obtaining good colour value and of avoiding colour bleeding, using alginates and other thickeners. Rompp, Axon and Thompson (1983), discuss the use of alginates with reactive dyes on cotton, viscose rayon and cotton-synthetic blend fabrics.

Ramakrishnan (1981) deals with the principles of reactive printing and the problems which arise in rotary printing machines, and the use of sodium alginate in the processing. Obenski (1984) has discussed the US printed fabric market, the use of alginates and guar gum as thickening agents and their share of the US market.

General reviews of thickeners in printing pastes, which include the use of alginates, have been made by Christie (1976), Shenai and Saraf (1981), Narkar (1982) and Teli, Shah and Sinha (1986).

Reviews dealing more specifically with the use of alginates in textile printing can be found in Ornaf (1969), Hilton (1972), Iwahashi (1975), Shah (1975), Balassa (1977), Khairoowala and Afrin (1984), Hebeish et al. (1986) and Teli and Chiplunkar (1986).

FOODS

Alginates have a long history of use in foods and these uses are based mainly on their thickening, gelling and general colloidal properties. Thickening is useful in sauces, syrups and toppings for ice cream, etc., pie fillings (it reduces moisture retention by the pastry), cake mixes (it thickens the batter aids moisture retention), and canned meat and vegetables (it can give either temporary or delayed-action thickening). Gel formation leads to uses in instant milk desserts and jellies, bakery filling cream, fruit pies, animal foods and reformed fruit. General colloidal properties are difficult to define but are illustrated by the results obtained by adding sodium alginate to ice cream and water ices, or propylene glycol alginate to stabilize beer foam or the suspended solids in fruit drinks (Leigh, 1979). Details of these and other applications can be found in some of the more recent reviews which have been written by Glicksman (1969a), McNeely and Pettitt (1973), McDowell (1975), Lawrence (1976), Cottrell and Kovacs (1980), Littlecott (1982), King (1983) and Sime (1984); their content is summarized below.

McNeely and Pettitt (1973) is well referenced to the general literature and much of the material is still useful. McDowell (1975) classifies uses according to the relevant alginate property such as thickening, gelling, film formation and stabilizing; it is a general review of value to a new user of alginates. Lawrence (1976) surveyed those US patents since the early 1960's that deal with gums for edible purposes and he includes a lengthy section on alginates. The four most recent reviews are all written by personnel from Kelco, the largest world manufacturer of alginate. Littlecott (1982) gives a good explanation of how to form food gels with alginates and provides many examples and formulations. Sime (1984) also deals with food gelling systems and is a useful addition to Littlecott's review; after discussing the general principles, he gives details for making reformed pimiento strips for olives, and structured fruits from fruit puree. Cottrell and Kovacs (1980) relate food applications to properties of alginate and give a wide variety of sample formulations for various food products; they give few literature references.

The review by King (1983) is excellent; he draws on the material used by Cottrell and Kovacs for describing the properties of alginates and then provides a thorough and well documented description of the food uses of alginates up to 1981-82.

One of the more recent developments is the use of alginates in restructured meat products. The US Department of Agriculture approved the use of alginate as a binder in these products last year (September, 1986) and this should lead to a new market for alginate. Restructuring is the process of taking flaked, sectioned or chunked meat and binding the pieces to resemble intact cuts of meat. The final products can be shaped as nuggets, roasts, loaves and steaks. Until now most restructured products have been sold frozen or cooked, so they could retain their shape. With the use of binders, the restructured products can be sold fresh or raw. The binder is a powder of sodium alginate, calcium carbonate, lactic acid and calcium lactate. When mixed with the raw meat, they form a calcium alginate gel which binds the meat. This binding mixture can be used to replace the sodium chloride and phosphate salts commonly used, thereby reducing the sodium level in the restructured products. Up to 1% sodium alginate is permitted. A patent has been assigned to the developers of the process, Colorado State University Research Foundation (Schmidt and Means, 1986), the inventors have published the information separately (Means and Schmidt, 1986) and the application is outlined by Andres (1987).

Alginates have been used for other re-formed foods. Morimoto (1984) patented a process for making shrimp or crabmeat analogue products using alginate and proteins such as soy protein concentrate or sodium caseinate. A mixture of the two is extruded into a clacium chloride bath to form edible fibres which are then frozen, thawed, chopped, coated with sodium alginate and formed in an appropriately shaped mould. After further freezing and thawing a product analogous to natural shrimp is obtained. Wylie (1976) described the manufacture of analogue fish fillets (sole) using minced white fish and a calcium alginate gel; the products could be grilled or cooked with sauce. A meat substitute has been formed from an aqueous mixture of protein and alginate by a process of freezing, slicing, gelling and heat setting; a well defined fibre structure results (Shenouda, 1983).

The principles used for making structured fruit products have been extended to making structured potato products such as croquettes and french fries (Anon., 1983). A synthetic potato skin shell containing alginate can be filled with mashed potato and browned to produce 'baked potatoes' in the fast-food market (Ooraikul and Aboagye, 1986). A patent by Cox (1982) for forming simulated, shaped, edible products includes the production of caviar and cottage cheese as examples.

There has been an increased interest in the use of alginate-pectin mixed gels with potential for use in jams, fruit flans and mayonnaise (Thom et al., 1982; Toft, 1982; Morris and Chilvers, 1984). New dessert gels from alginate have been reviewed by Kelco (1983) while Protan (1986a) has discussed alginates as stabilizers in bakery creams, jams and jellies.

The reasons for the effectiveness of alginate as a stabilizer in ice cream have never been fully understood; Muhr and Blanshard (1984) have studied the mechanism for the reduction of crystal growth but their work is not yet conclusive.

A moisture barrier which allows breaded or batter-covered products to come in contact with a sauce or filling can be made using a coating of soluble alginate which is then treated with calcium chloride (Earle and McKee, 1986). Alginates are being used to make improved rice pasta and vegetable pasta (Hsu, 1985, 1985a). Calcium alginate can be formed as a fibrous precipitate and is used to simulate the texture of natural fruit and vegetables (Anon., 1980).

IMMOBILIZED BIOCATALYSTS

Many commercial chemical syntheses and conversions are best carried out using biocatalysts such as enzymes or whole cells. Examples are (a) the use of enzymes for the conversion of glucose (40% of the sweetness of sucrose) to fructose (about 150% of the sweetness of sucrose), the production of L-amino acids for use in foodstuffs, the synthesis of new penicillins after hydrolysis of penicillin G, (b) the use of whole cells to promote the conversion of starch to ethanol, for beer brewing, for the continuous production of yoghurt. To carry out such processes on a moderate to large scale, the biocatalysts need to be in a concentrated form and to be recoverable from the process for reuse. This can be achieved by "immobilizing" the enzymes or cells; they can be fixed to the surface of an insoluble solid or entrapped in a polymeric material. In the 1970s many single enzymes were isolated, immobilized and used, but more recently it has been found that it is easier, more economical and often more effective to immobilize whole cells, which contain multi-enzyme systems. An added advantage of immobilizing cells is the increased stability often found; it is not unusual for a half-life of one day for ordinary suspended cells, to be increased to 30 days for immobilized or resting cells. A good introduction to the reasons for using biocatalysts and for their immobilization is given by Tramper (1985); he also describes the use of alginate for immobilization. Bucke and Wiseman (1981) give a more detailed background and review the developments to the early 1980s.

Alginate gels have proved to be a very successful medium for entrapping biocatalysts, especially when formed as beads of gel. The cell suspension is mixed with sodium alginate solution (2-4%) and this is extruded as drops into calcium chloride solution (0.05-0.1M). An immediate skin forms around the drop and as calcium ions gradually diffuse inwards, a gel forms. The size of the beads can be regulated from the size of the needle or nozzle, usually 0.2-1.0 mm but up to 5 mm. The fresh beads can be separated and used or they can be dried; drying increases their strength and reduces their ability to swell so they contain, when rewetted, more cells per unit volume.

Alginates which form strong gels are best for this application; the alginate should therefore contain a high proportion of guluronic acid, such as that extracted from the stipes of Laminaria hyperborea, a species particularly abundant in the cold waters of Norway. There have been recent reports of high guluronic acid contents in seaweeds from warmer waters, Sargassum species from Sabah, Malaysia (Wedlock, Fasihuddin and Phillips, 1986) and Sargassum, Turbinaria and Cystoseira from Sri Lanka (Shyamali, de Silva and Savitri Kumar, 1984).

Details of methods are available from manufacturers (such as Protan, 1987) or from the literature (Dallyn, Falloon and Bean, 1977; Klein and Wagner, 1982; Klein, Stock and Vorlop, 1983; Tanaka, Matsumura and Veliky, 1984; Rehg, Dorger and Chau, 1986).

Johansen and Flink (1985, 1986, 1986a, 1986b) have applied the internal gelation principles, developed for food gels, to immobilization techniques using yeast cells for their studies. Sodium alginate, an insoluble calcium salt and D-glucono-1,5-lactone are mixed in water; the lactone slowly hydrolyses, lowers the pH, releases calcium ions and gelation occurs gradually, from within the solution. The resulting immobilizates have particles of higher strength, with at least equal fermentation rates, when compared to externally gelled material. Rochefort, Rehg and Chau (1986) have stabilized calcium alginate gels by washing with 0.1M aluminium nitrate. Burns, Kvesitadze and Graves (1985), produced dried spheres of calcium alginate containing magnetite and found they have good potential as a support for enzyme immobilization.

Cell immobilization with alginate can be done under mild conditions with little loss of activity of the cells and the activity is often stable for extended periods of time. Temperatures can be 0-100°C and the pH neutral but any buffers used must not contain citrate or phosphate. These anions will remove calcium ions from the gel and can lead to its breakdown, although Birnbaum et al. (1981), have developed methods for stabilizing alginate gels in phosphate-containing media. The cell-gel entrapment can be done under sterile conditions and the alginate gel is stable (0-100°C) and non-toxic. The cells can be recovered if necessary be adding a sequestering agent for the calcium ions (such as polyphosphate or EDTA); once the calcium ions are removed from the gel, its structure is lost and it changes to a liquid with the cells suspended in it.

The number of processes in which alginate has been used for cell or enzyme immobilization, on laboratory and larger scales, has increased dramatically in the last few years. Good sources of these works are journals such as Biotechnology Letters, Biotechnology and Bioengineering or the abstracts available from data bases such as Biobusiness (Dialog Information Services) and Current Biotechnology Abstracts (Pergamon Infoline). Some examples are:

(a) production of ethanol from starch (McGhee, Carr and St. Julian, 1984);

(b) beer brewing with immobilized yeast (Onaka et al., 1985);

(c) production of citric acid (Lim and Choi, 1986);

(d) continuous yoghurt production (Prevost, Divies and Rousseau, 1985);

(e) fermentation to produce butanol and isopropanol (Schoutens et al., 1986);

(f) continuous acetone-butanol production (Frick and Schuegerl, 1986);

(g) pilot-plant production of prednisolone from hydrocortisone (Kloosterman and Lilly, 1986);

(h) glycerol production from the marine alga, Dunaliella tertiolecta (Grizeau and Navarro, 1986).

PAPER

Until the late 1950s the main use for alginate in the paper industry was in surface sizing. Its addition to the normal starch sizing gives a smooth continuous film and a surface with less fluffing. The oil resistance of alginate films give a size with better oil resistance so an improved gloss is obtained with high gloss inks. If papers or boards are to be waxed, alginate in the size will keep the wax mainly at the surface. The quantity of alginate used is usually 5-10% of the weight of starch in the size.

Alginate is also used in starch adhesives for making corrugated boards because it stabilizes the viscosity of the adhesive and allows control of its rate of penetration. One percent sodium alginate, based on the weight of starch used, is usually sufficient.

Cottrell and Kovacs (1980) give examples of formulations for a kraft lineboard sizing and for corrugating adhesives. An improved sizing with alginate has been obtained by using a paper containing 5-25% of calcium carbonate filler; the calcium alginate film which forms gives better solvent resistance and lower Bendtsen porosity. The alginate is blended with 6-20 parts of starch or it may be combined with polyvinyl alcohol (Kelco, 1985).

Paper coating methods and equipment have developed significantly since the late 1950s as the demand for a moderately priced coated paper for high quality printing. Trailing blade coating equipment runs at 1 000 metres per minute or more so the coating material, usually clay plus a synthetic latex binder, must have consistent rheological properties under the conditions of coating. Up to 1% alginate will prevent change in viscosity of the coating suspension under the high shear conditions where it contacts the roller. The alginate also helps to control water loss from the coating suspension into the paper, between the point where the coating is applied and the point where the excess is removed by the trailing blade. The viscosity of the coating suspension must not be allowed to increase by loss of water into the paper because this leads to uneven removal by the trailing blade and streaking of the coating. Medium to high viscosity alginates are used, at a rate of 0.4-0.8% of the clay solids. A new modified form of sodium alginate has been reported to be more effective than existing alginate and results in lower processing costs (Yin and Grishaber, 1979).

A useful discussion of the evolution of paper coating methods and the use of alginate has been prepared by Blood (1968). Other more general discussions of coating, which also refer to the use of alginates, have been published by Paper (Anon., 1976), Bergmann and Hunger (1978), and Delaplace, Laraillet and Isoard (1985).

Sergeant (1981) has prepared a general review of the applications of alginates in paper converting. He has also described the use of zinc ammonium alginate as a flame retardant in paper (Sergeant, 1980).

WELDING RODS

Coatings are applied to welding rods or electrodes to act as a flux and to control the conditions in the immediate vicinity of the weld, such as temperature or oxygen and hydrogen availability. The dry ingredients of the coating are mixed with sodium silicate (water glass) which gives some of the plasticity necessary for extrusion of the coating onto the rod and which also acts as the binder for the dried coating on the rod. However the wet silicate has no binding action and does not provide sufficient lubrication to allow effective and smooth extrusion. An additional lubricant is needed, and a binder which will hold the damp mass together before extrusion and maintain the shape of the coating on the rod during drying and baking. Alginates are used to meet these requirements.

Soluble alginates (sodium or potassium) are used in coatings on welding rods which are dried at moderate temperatures and in which the alginate remains after drying; this includes organic-type coatings with a high content of cellulosic material and mineral coatings of the "acid" type. Soluble alginates can be used in basic or low-hydrogen rods but calcium alginate, sometimes with a proportion of sodium alginate added, gives much better results. This is related to the high temperatures used to dry these rods (400-450°C) which produces low moisture contents so that only very low hydrogen levels are found in the deposited weld metals. Soluble alginates swell when wetted and as the water is driven out completely in this high-temperature drying, the alginates will contract and cracks will develop in the coating. When calcium alginate is mixed with sodium silicate, a small amount of sodium alginate forms around each particle of calcium alginate. This mixture is thixotropic, its viscosity is lowered when extrusion pressure is applied; it therefore acts as a good binder and extrusion lubricant. During the drying process, because the calcium alginate did not previously swell very much, it does not shrink appreciably and a more uniform coating results.

The quantities of alginates used are very dependent on the type of welding rod being coated and the extrusion equipment being used. For soluble alginates it may be 0.4-1.2% for low-hydrogen welding rods and 0.15-0.25% for acid and organic types. For the thixotropic alginates, manufacturers often find it more effective to use a mixture of calcium alginate and sodium alginate with a total alginate content of 0.4-0.6% for low hydrogen electrodes. Alginate manufacturers are the best source of information for using alginates in welding rod applications, for example Protan (1984).

PHARMACEUTICAL

Alginic acid is insoluble in water but swells when placed in water. This property makes it a useful disintegrating agent in tablets. It is more expensive than the traditional disintegrating agent, starch, but its overall addition to the cost of the tablet is still usually very low. It is a better disintegrant than starch so less is required. It can be added during the granulating process, rather than as a powder after granulation, so the processing is easier. The mechanical strength of the final tablet is greater, compared to using starch.

Sodium alginate is used in some liquid medicines to increase viscosity and improve the suspension of solids. Propylene glycol alginate can improve the stability of emulsions. Capsules containing sodium alginate and calcium carbonate are used to protect inflamed areas near the entrance to the stomach. The acidity of the stomach causes formation of insoluble alginic acid and carbon dioxide; the alginic acid rises to the top of the stomach contents and forms a protective layer.

Very useful dental impression compounds are based on alginate cold-setting gels; some recent examples can be found in Pellico (1983) and Scheuble and Munsch (1983). Alginates are the basis of many slimming or diet foods, particularly biscuits; alginic acid swells in the stomach and fills it so that the dieter no longer feels hungry; the body cannot assimilate the alginic acid so no calories are absorbed.

These uses have been discussed in earlier reviews such as McNeely and Pettitt (1973).

OTHER USES

MEDICAL DRESSINGS

Courtaulds (UK) has patented a wound dressing which combines aspects of alginate filament formation with those of spunbonding to produce a good quality staple fibre (Aldred and Mosely, 1983). This fibre can be easily processed into nonwoven fabrics. The sodium calcium alginate fibres are useful as haemostatic wound dressings which can be absorbed by body fluids, as the calcium in the fibre is exchanged for sodium from the body fluids (Burrow and Welch, 1983).

A new "biopaper" has been made by the Japan Institute of Industrial Research (1985). The papers, made from alginic acid or a mixture of the acid and its calcium salt, are expected to be of value for bandages and similar medical uses where the haemostatic properties of alginates are useful. Bioactive papers have been made from staple fibres in which enzymes have been entrapped (Kobayashi, 1986; Kobayashi and Matsuo, 1986; Kobayashi, Matsuo and Kawakatsu, 1986).

CONTROLLED RELEASE OF CHEMICALS

This use has some similarities to the methods used for immobilization of cells. In this application the cells in the alginate gel beads are replaced by materials having biological activity, such as biological and chemical herbicides. The rate of release of the herbicides into soil or water can be controlled by the properties of the gel beads; the beads can be air dried and become hard granules. By incorporating air into the beads they can be made to float. The patent suggests their use for herbicides, pesticides, algicides and most biologically active substances (Connick, 1983; Connick, Lee and Rawson, 1984).

An alginate-clay mixture and calcium ions have been used to encapsulate microorganisms (chiefly fungi) which have potential to control plant diseases (Fravel et al., 1985). A sustained release system for pharmaceuticals, using calcium alginate beads, has been reported by Badwan et al. (1985).

BINDERS FOR FISH FEEDS

The worldwide growth in aquaculture has led Protan (1978) to investigate the use of alginate as a binder in fish feeds, especially moist feed made from fresh waste fish mixed with various dry components. Alginate binding can lower consumption by up to 40% and pollution of culture ponds is sharply reduced. More recent technical information is available from the authors.

CONFECTIONERY

Alginate gels find a small application in confectionery. Recently the incorporation of fruit pulp has been suggested and a method for making Turkish delight is described (Anon., 1983a).

RELEASE AGENTS

The poor adhesion of films of alginate to many surfaces, together with their insolubility in non-aqueous solvents, have led to their use as mould release agents, originally for plaster moulds and later in the forming of fibreglass plastics. Sodium alginate also makes a good coating for anti-tack paper which is used as a release agent in the manufacture of synthetic resin decorative boards (Cheetham, 1976; Sumitomo Bakelite, 1981). Films of calcium alginate, formed in situ on a paper, have been used to separate decorative laminates after they have been formed in a hot-pressing system (Jaisle and Bunkowski, 1981; Jaisle and Schiermeier, 1981).

MARKETING

There are difficulties and costs in the marketing of seaweed colloids, such as alginate, agar and carrageenan, which are not always apparent to those outside the industry. In some markets one colloid may compete with another, in others one might be the only real choice. They must all compete, in at least some of their uses, with plant gums (such as guar and locust bean) and cellulose derivatives (such as CMC and methyl cellulose) which are often cheaper. It is important to realize that price may not be the determining factor in a buyer's choice of a seaweed colloid; quality and its reproducibility from one batch to another may be more important. Frequently a buyer uses less than 1% of the colloid in his product so a 20% price difference may be inconsequential in the total cost of his product. Many a buyer of seaweed colloids, satisfied with one particular brand or grade, will, despite a higher price, stay with it because the risks of changing may not seem to be worth the saving. So in seaweed colloids, those brands already established in the market often hold a very strong, entrenched position. To dislodge them, a marketing group should include a strong technical team which can run tests and trials to convince the buyer of the equivalence of the new product; sometimes this requires a detailed knowledge of the buyer's industry. In promoting new sales, the colloid producer may have to provide complete formulations and technical know-how to potential buyers. Therefore selling costs of the seaweed colloids can be high and account must be taken of this by the potential producer.

The buyers of alginates fall into two groups. The first is a number of large buyers who know exactly what they want and who require little servicing because they have their own resources. This group includes those specialty gum companies who service smaller users by preparing their own blends of seaweed colloids and other colloids, according to the requirements of a particular customer.

The second and larger group are the smaller users who need some technical service support. Frequently this group yields more profitable sales in the long term because they may be sold specifically formulated products at a premium price and they are generally more reluctant to change to a competitor's product. On the other hand it takes more time and expense to establish such sales. Many of the major producers have such specialty products, shown by the large range of products listed by them.

Alginate manufacturers usually sell direct to the major markets but in minor markets it is more economical to sell through an agent, leaving the task of market penetration to him but providing technical support where necessary. Agents need to have an appreciation of the application of colloids and a knowledge of the client's industry. This ideal might be achieved by an agent selling a variety of chemicals to just one industry, like the food industry, but an agent who deals principally in colloids over a range of industries, which is a not uncommon situation, usually needs more backup from the producer. Large wholesalers/agents may buy from the producer and resell; otherwise they operate on a commission of 5-15%, depending partly on the degree of assistance required from the producer.

The world market for alginates is estimated at 20 000-24 000 tonnes per year (Kjemi, February 1986; Inf.Chim., (265), October 1985) which is similar to estimates made in 1980 (ITC, 1981). Demand for sodium alginate is steady with plentiful supplies.

Three principal grades are available but there are variations of viscosity, and many specially formulated mixtures with additives, within each grade. The highest grades meet the requirements of the National Formulary (USA), food grades generally meet the quality standards of the Food Chemicals Codex (USA) and technical grades vary considerably in their colour and water-insoluble solids (such as cellulose). Other countries or groups (EEC) have similar specifications to the NF and FCC.

Prices have shown little change between 1986 and 1987 and lie in the following ranges: Sodium alginate: pharmaceutical (NF) grade, US$ 6-7 per pound; food (FCC) grade, US$ 3-5 per pound; technical grade, US$ 2.50-3.50 per pound. Propylene glycol alginate, US$ 6-7 per pound (Chemical Marketing Reporter, 27 January 1986, 11 August 1986, 23 February 1987).

REFERENCES

Aldred, F.C. and C.R. Mosely, 1983. Man-made filaments and methods of making wound dressings containing them. U.S. Patent 4,421,583.

Andres, C., 1987. Expanding applications for alginate technologies. Food Process., 48(2):30-2

Annison, G., N.W.H. Cheetham and I. Couperwhite, 1983. Determination of the uronic acid composition of alginates by high-performance liquid chromatography. J.Chromatogr., 204:137-43

Atkins, E.D.T., et al., 1973. Structural components of alginic acid. 1. Crystalline structure of poly-β-D-mannuronic acid. Results of x-ray diffraction and polarized infrared studies. Biopolymers, 12:1865-78

_____, 1973a. Structural components of alginic acid. 2. Crystalline structure of poly-α-L-guluronic acid. Results of x-ray diffraction and polarized infrared studies. Biopolymers, 12:1879-87

Badwan, A.A., et al., 1985. A sustained release drug delivery system using calcium alginate beads. Drug Dev.Ind.Pharm., 11:239-56

Balassa, A., 1977. Alginates from Chilean seaweed. Melliand Textilber.Int/Int.Textile Rep.(Eng.Ed.), 6:117-8

Bergmann, W. and G. Hunger, 1978. Present status and trends in coated paper and board grades. Wochenbl.Papierfabr., 106(18): 703-8 (in German)

Birnbaum, S., et al., 1981. Covalent stabilization of alginate gel for the entrapment of living whole-cells. Biotechnol. Lett., 3:393-400

Black, W.A.P. and F.N. Woodward, 1954. Alginates from common British brown marine algae. In Natural plant hydrocolloids. Adv.Chem.Ser.Am.Chem.Soc., 11:83-91

Blood, C.T., 1968. Sodium alginate in modern paper coating. London, Alginate Industries Ltd., 22 p.

Booth, E., 1975. Seaweeds in industry. In Chemical oceanography, edited by J.P. Riley and G. Skirrow, New York, Academic Press, Vol.4:219-68

Braud, J.-P., R. Debroise and R. Pérez, 1977. Nouvelles perspectives dans l'exploitation des laminaires (New prospects for utilization of marine algae of the genus Laminaria). Rev.Trav.Inst.Pêches Marit.,Nantes, 41:203-11 (in French)

Bucke, C. and A. Wiseman, 1981. Immobilized enzymes and cells. Chem.Ind,, 4 April issue:234-40

Burns, M.A., G.I. Kvesitadze and D.J. Graves, 1985. Dried calcium alginate/magnetite spheres: a new support for chromatographic separations and enzyme immobilization. Biotechnol.Bioeng., 27(2):137-45

Burrow, T.R. and M.J. Welch, 1983. The development and use of alginate fibre in nonwovens for medical end-uses. In Nonwovens Conference papers. Manchester, U.K. University of Manchester Institute of Science and Technology (UMIST), pp.49-68

Cheetham, J.R., 1976. High pressure laminate. German Patent 2,551,577

Christie, N.J., 1976. Thickeners for Thermosol and space-dyeing. Int.Dyer, 155:19-23

Clark, D.E. and H.C. Green, 1936. Alginic acid and process of making same. U.S. Patent 2,036,922

Connick, W.J. Jr., 1983. Controlled release of bioactive materials using alginate gel beads. U.S. Patent 4,401,456

Connick, W.J. Jr., R.E. Lee and J. Rawson, 1984. Encapsulation with seaweed-based gels: a new process. Agric.Res., 32(10):8-9

Cottrell, I.W. and P. Kovacs, 1980. Alginates. In Handbook of water-soluble gums and resins, edited by R.L. Davidson. New York, McGraw-Hill, pp.2.1 to 2.43

Cox, J.P., 1982. Method for forming shaped products for human and/or animal consumption or as marine bait and products produced thereby. U.S. Patent 4,362,748

Dallyn, H., W.C. Falloon and P.G. Bean, 1977. Method for the immobilization of bacterial spores in alginate gel. Lab.Pract., 26:773-5

Delaplace, P., P. Laraillet and J.C. Isoard, 1985. Coating binders: update. ATIP(Assoc.Tech.Ind.Papet.)Rev., 39(2):95-100 (in French)

Earle, R.D. and D.H. McKee, 1986. Coated food product and method of making same. U.K. Patent Appl. 2,172,183A

Fravel, D.R., et al., 1985. Encapsulation of potential biocontrol agents in an alginate-clay matrix. Phytopathology, 75:774-7

Frick, C. and K. Schuegerl, 1986. Continuous acetone-butanol production with free and immobilized Clostridium acetobutylicum. Appl.Microbiol.Biotechnol., 25:186-93

Glicksman, M., 1969. Gum technology in the food industry. New York, Academic Press, pp.67-8

_____, 1969a. Gum technology in the food industry. New York, Academic Press, pp.239-73

Grant, G.T., et al., 1973. Biological interactions between polysaccharides and divalent cations: the egg-box model. FEBS Lett., 32:195-8

Grasdalen, H., B. Larsen and O. Smidsrod, 1970. A.P.M.R. study of the composition and sequence of uronate residues in alginates. Carbohydr.Res., 68:23-31

_____, 1981. ^{13}C-NMR studies of monomieric composition and sequence in alginate. Carbohydr.Res., 89:179-91

Green, H.C., 1936. Process for making alginic acid and product. U.S. Patent 2,036,934

Grizeau, D. and J.M. Navarro, 1986. Glycerol production by Dunaliella tertiolecta immobilized with calcium alginate beads. Biotechnol.Lett., 8:261-4

Haug, A., 1964. Composition and properties of alginates. Rep.Norw. Inst.Seaweed Res., (30):123 p.

Haug, A. and B. Larsen, 1962. Quantitative determination of the uronic acid composition of alginates. Acta Chem.Scand., 16:1908-18

Haug, A., B. Larsen and O. Smidsrod, 1966. A study of the constitution of alginic acid by partial acid hydrolysis. Acta Chem.Scand., 20:183-90

_____, 1974. Uronic acid sequence in alginate from different sources. Carbohydr.Res., 32:217-25

Haug, A., et al., 1967. Correlation between chemical structure and physical properties. Acta Chem.Scand., 21:768-78

Hebeish, A., et al., 1986. Technical feasibility of some thickeners in printing cotton with reactive dyes. Am.Dyestuff Report., 75(2):22-9,43

Henkel et Cie., 1964. G.m.b.H. Bleaching agents for alginic acid and derivatives. French Patent 1,360,504

Hilton, K.A., 1969. Alginates in textile printing. In CIBA review. Basel, Switzerland, CIBA, pp.19-46

_____, 1972. Sodium alginate in the textile industry. Colourage, 19(10):65-8

Hsu, J.Y., 1985. Rice pasta composition. European Patent 0,081,077 B1

_____, 1985a. Preparation of vegetable pastas. U.S. Patent 4,517,215

Ishikawa, M., 1984. Unpublished observation at the Division of Marine Radioecology, National Institute of Radiological Science, Isozaki, Nakaminato, Japan

ITC (International Trade Centre), 1981. Pilot survey of the world seaweed industry and trade. Geneva, UNCTAD/GATT, 111 p.

Iwahashi, A., 1975. Sodium alginate as a textile printing thickener. Japan Textile News, (252):97-100

Jaisle, R.F. and K.D. Bunkowski, 1981. Process for releasing laminates. U.S. Patent 4,243,461

Jaisle, R.F. and W.W. Schiermeier, 1981. Process for releasing laminates. U.S. Patent 4,263,073

Japan Frozen Foods Inspection Corporation, 1984. Kobe Branch, Kobe. Examination certificate, K 59, 101, No. 117 (in Japanese)

Japan Institute of Industrial Research, 1985. News item. Jap.Chem.Week, 26(1329)3. Issued also in Japan Econ.J., 23(1176)16

Ji, M.H., et al., 1984. Studies on the M:G ratios in alginate. Hydrobiologia, 116/117:554-6

Johansen, A. and J.M. Flink, 1985. A novel method for immobilization of yeast cells in alginate gels of various shapes by internal liberation of calcium ions. Biotechnol.Lett., 7(10):765-8

_____, 1986. A new principle for immobilized yeast reactors based on internal gelation of alginate. Biotechnol.Lett., 8(2):121-6

_____, 1986a. Immobilization of yeast cells by internal gelation of alginate. Enzyme Microbial Technol., 8(3):145-8

_____, 1986b. Influence of alginate properties and gel reinforcement on fermentation characteristics of immobilized yeast cells. Enzyme Microbial Technol., 8(12):737-48

Kelco, 1952. Alkylene glycol esters of alginic acid. British Patent 676,618

_____, 1976. Kelco algin, hydrophilic derivatives of alginic acid for scientific water control. San Diego, Delco Division of Merck and Co. Inc., 51 p. 2nd ed.

_____, 1983. New dessert gels. Chilton's Food Eng., September issue: 66-7

_____, 1985. Surface sizing of paper using sodium alginate in the presence of calcium carbonate fillers. Res.Disclosure, (257):September issue: 435

Khairoowala, Z.U. and S. Afrin, 1984. Sodium alginate - an assessment of scope. Textile Ind.Trade J., 22(5/6):9-12

Kim, Dong-Soo, 1984. Uronic acid composition, block structure and some related properties of alginic acid. Master's thesis. Pusan Commercial (San Eup) University, Pusan, Republic of Korea

King, A.H., 1983. Brown seaweed extracts (alginates). In Food hydrocolloids, edited by M. Glicksman. Boca Raton, Florida, CRC Press, pp.115-88

Klein, J. and F. Wagner, 1982. Preparation of immobilized enzymatically-active substance (on alginate gels). U.S. Patent 4,334,027

Klein, J., J. Stock and K. Vorlop, 1983. Pore size and properties of spherical calcium-alginate biocatalysts. Eur.J.Appl. Microbiol.Biotechnol., 18:86-91

Kloosterman, J. and M.D. Lilly, 1986. Pilot-plant production of prednisolone using calcium alginate immobilized Arthrobacter simplex. Biotechnol.Bioeng., 28:1390-5

Kobayashi, Y., 1986. Papers from seaweeds. Manufacture and applications of alginate fiber papers. Tanpakushitsu Kakusan Koso, 31:1066-77 (in Japanese)

Kobayashi, Y. and R. Matsuo, 1986. Manufacture of alginate paper. Japanese Patent Kokai 174,499/86

Kobayashi, Y., R. Matsuo and H. Kawakatsu, 1986. Manufacture and physical properties of alginate fiber papers as an analysis model of cellulosic fiber papers. J.Appl.Polymer Sci., 31:1735-47

Larsen, B., 1981. Biosynthesis of alginate. Proc.Int.Seaweed Symp., 10:7-34

Larsen, B. and A. Haug, 1971. Biosynthesis of alginate. Part 2. Polymannuronic acid C-5-epimerase from Azobacter vinelandii (Lipman). Carbohydr.Res., 17:297-308

Lawrence, A.A., 1976. Natural gums for edible purposes. Park Ridge, New Jersey, Noyes Data Corp., pp.132-86

Le Gloahec, V.C.E., 1939. Fixation of chlorophyllian colored matter. U.S. Patent 2,163,147.

Le Gloahec, V.C.E. and J.B. Herter, 1938. Method of treating seaweed. U.S. Patent 2,128,551.

Leigh, A.M., 1979. Alginates in food production. London, Alginate Industries Ltd., 11 p.

Lim, D.J. and C.Y. Choi, 1986. Citric acid production using immobilized yeast activated with calcium chloride-containing medium. Korean J.Appl.Microbiol. Bioeng., 14:285-92 (in Korean)

Littlecott, G.W., 1982. Food gels - the role of alginates. Food Technol.Aust., 34:412-8

Maass, H., 1959. Alginsäure und alginate. Heidelberg, Strassenbau chimie und Technik Verlagsgesellschaft m.b.H.

McDowell, R.H., 1975. New developments in the chemistry of alginates and their use in food. Chem.Ind., 1975:391-5

──────────, 1960. Applications of alginates. Rev.Pure Appl.Chem., 10:1-19

──────────, 1977. Properties of alginates. London, Alginate Industries Ltd., 67 p. 4th ed.

McGhee, J.E., M.E. Carr and G. St. Julian, 1984. Continuous bioconversion of starch to ethanol by calcium alginate-immobilized enzymes and yeasts. Cereal Chem., 61:446-9

McNeely, W.H. and D.J. Pettitt, 1973. Algin. In Industrial gums, edited by R.L. Whistler. New York, Academic Press, pp.49-81. 2nd ed.

Means, W.J. and G.R. Schmidt, 1986. Algin/calcium gel as a raw and cooked binder in structured beef steaks. Food Sci., 51:60-5

Morimoto, K., 1985. Extrusion process for shrimp or crabmeat analog products in a series of non-boiling gelling baths. U.S. Patent 4,554,166

Morris, E.R., D.A. Rees and D. Thom, 1980. Characteristics of alginate composition and block-structure by circular dichroism. Carbohydr.Res., 81:305-14

Morris, V.J., 1985. Food gels - roles played by polysaccharides. Chem.Ind., 4 March issue: 159-64

Morris, V.J. and G.R. Chilvers, 1984. Cold setting alginate-pectin mixed gels. J.Sci.Food Agric., 35:1370-6

Moss, J.R. and M.S. Doty, 1987. Establishing a seaweed industry in Hawaii: an initial assessment. Honolulu, Hawaii State Department of Land and Natural Resources. 73 p.

Muhr, A.H. and J.M.V. Blanshard, 1984. The effect of polysaccharide stabilizers on ice crystal formation. In Gums and stabilizers for the food industry, edited by G.O. Phillips, D.J. Wedlock and P.A. Williams. Oxford, Pergamon, Vol.2:327-31

Narkar, A.K., 1982. Thickeners for printing Reactive dyes. Indian Textile J., 92(7):117-22

Noto, V.H. and D.J. Pettitt, 1980. Production of propylene glycol alginic acid esters. British Patent 1,563,019.

Obenski, B.J. 1984. Printing and print paste thickeners. Am.Dyestuff Report., 73:15-6,39

Okazaki, A., 1971. Seaweeds and their uses in Japan. Tokyo, Tokai University Press, 165 p.

Onaka, T., et al., 1985. Beer brewing with immobilized yeast. Bio/Technology, 3:467-70

Ooraikul, B. and N.Y. Aboagye, 1986. Synthetic food product. U.S. Patent 4,582,710

Ornaf, W., 1969. Sodium alginate - relations between structural properties and application for (textile) printing. Textilveredlung, 4:850-61 (in German)

Pellico, M., 1983. Settable alginate compositions. U.S. Patent 4,381,947

Penman, A. and G.R. Sanderson, 1972. A method for the determination of uronic acid sequence in alginates. Carbohydr.Res., 25:280

Pettitt, D.J. and V.H. Noto, 1973. Process for the preparation of propylene glycol alginate from partially neutralized alginic acid. U.S. Patent 3,772,266

Prelini, G., 1982. Printing of cotton with reactive dyes. Textilia, 58:31-2 (in Italian)

Prevost, H., C. Divies and E. Rousseau, 1985. Continuous yoghurt production with Lactobacillus bulgaricus and Streptococcus thermophilus entrapped in calcium alginate. Biotechnol. Lett., 7:247-52

Protan, 1984. Alginates in welding rod coverings. Drammen, Norway, Protan A/S, 10 p.

_____, 1985. Protan textile alginates. Drammen, Norway, Protan A/S, 20 p.

_____, 1986. Alginate technical information bulletin. Drammen, Norway, Protan A/S, 15 p.

_____, 1986a. Alginates. Chilton's Food Eng.Int., 11(2):60

Protan, 1987. Protanal alginates for cell immobilization. Drammen, Norway, Protan A/S, 3 p.

Protan and A.S. Fagertun, 1978. Protanal alginates as binders in fish feeds. Riv.Ital.Piscic.Ittiopatol., 13(4):103-6 (in English)

Racciato, J.S., 1979. Printing cationic dyes with xanthan gum or algin. Textile Chem.Colorist, 11:46-50

Ramakrishnan, S., 1981. Textile thickening agents for reactive printing. Colourage, 28:9-11

Rees, D.A., 1969. Structure, conformation and mechanism in the formation of polysaccharide gels and networks. Adv.Carbohydr.Chem.Biochem., 24:303-4

_____, 1972. Polysaccharide gels, a molecular view. Chem.Ind., 19 Aug. issue: 630-6

Rehg, T., C. Dorger and P.C. Chau, 1986. Application of an atomiser producing small alginate gel beads for cell immobilization. Biotechnol.Lett., 8(2):111-4

Rochefort, W.E., T. Rehg and P.C. Chau, 1986. Trivalent cation stabilization of alginate gel for cell immobilization. Biotechnol.Lett., 8(2):115-20

Rompp, W., G.L. Axon and T. Thompson, 1983. Sodium alginate: a textile printing thickener. Am.Dyestuff Report., 72:31-2

Scheuble, M. and P. Munsch, 1983. Non-dusting and fast-wetting dental impression material. U.S. Patent 4,394,172

Schmidt, G.R. and W.J. Means, 1986. Process for preparing algin/calcium gel structured meat products. U.S. Patent 4,603,054

Schoutens, G.H., et al., 1986. A comparative study of a fluidized bed reactor and a gas lift loop reactor for the IBE process. Part 3. Reactor performances and scale up. J.Chem.Technol.Biotechnol., 36:565-76

Sergeant, S.V., 1980. Improving the flame resistance of paper. Converter, 17(10):14-5

_____, 1981. Applications of alginates in paper. Converter, 18(6):28-9

Shah, A.C., 1975. Chemistry and application of sodium alginate. Man-Made Text.India, 18:681-5, 687

Shenai, V.A. and N.M. Saraf, 1981. Thickeners and additives in the printing paste. Colourage, 28(22):44-8 Special supplement on textile printing

Shenouda, S.Y.K., 1983. Fabricated protein fibre bundles. U.S. Patent 4,423,083

Shyamali, S., M. de Silva and N. Savitri Kumar, 1984. Carbohydrate constituents of the marine algae of Sri Lanka. 2. Composition and sequence of uronate residues in alginates from some brown seaweeds. J.Nat.Sci.Counc.Sri Lanka, 12:161-6

Sime, W.J., 1984. The practical utilization of alginates in food gelling systems. In Gums and stabilizers for the food industry, edited by G.O. Phillips, P.J. Wedlock and P.A. Williams, Oxford, Pergamon, Vol.2:177-88

Skjak-Braek, G., 1984. Enzymatic modification of alginate. In Gums and stabilizers for the food industry, edited by G.O. Phillips, D.J. Wedlock and R.A. Williams, Oxford, Pergamon, Vol.2:523-8

Smidsrod, O. and A. Haug, 1972. Dependence upon the gel-sol state of the ion-exchange properties of alginates. Acta Chem. Scand., 26:2063-74

Smidsrod, O., A. Haug and S.G. Whittington, 1972. The molecular basis for some physical properties of polyuronides. Acta Chem.Scand., 26:2563-4

Stanford, E.C.C., 1881. Improvements in the manufacture of useful products from seaweeds. British Patent 142

Steiner, A.B., 1947. Manufacture of glycol alginates. U.S. Patent 2,426,215.

Steiner, A.B. and W.H. McNeely, 1950. High-stability glycol alginates and their manufacture. U.S. Patent 2,494,911.

_____, 1954. Algin in review. Adv.Chem.Ser.Am.Chem.Soc., 11:68-82

Sumitomo Bakelite Co. Ltd., 1981. Synthetic resin decorative boards. Japanese Patent Kokai 106,821/81

Tanaka, H., M. Matsumura and I.A. Veliky, 1984. Diffusion characteristics of substrates in calcium alginate gel beads. Biotechnol.Bioeng., 26:53-8

Teli, M.D. and V. Chiplunkar, 1986. Role of thickeners in final performance of reactive prints. Textile Dyer Printer, 19(6):13-9

Teli, M.D., R. Shah and R. Sinha, 1986. Developments in thickeners for reactive printing. Textile Dyer Printer, 19(7):17-23

Thom, D., et al., 1982. Interchain associations of alginates and pectins. Progr.Food.Nutr.Sci., 6:97-108

Thornley, F.C. and M.J. Walsh, 1931. Process of preparing alginic acid and compounds thereof. U.S. Patent 1,814,981

Toft, K., 1982. Interactions between pectins and alginates. Progr. Food Nutr.Sci., 6:89-96

Tramper, J., 1985. Immobilizing biocatalysis for use in synthesis. Trends Biotechnol., 3:45-9

Wedlock, D.J., B.A. Fasihuddin and G.O. Phillips, 1986. Characterization of alginates from Malaysia. In Gums and stabilizers for the food industry, edited by G.O. Phillips, D.J. Wedlock and P.A. Williams. London, Elsevier, Vol.3:47-67

Wylie, A., 1976. Alginates in the processing of minced fish. In The production and utilization of mechanically recovered fish flesh, Conference proceedings, edited by J.N. Keag, Aberdeen, Ministry of Agriculture, Fisheries and Food, Torry Research Station, pp. 87-92

Yin, R.I. and R. Grishaber, 1979. Paper coating trials of a new modified sodium alginate product. In TAPPI Coating Conference Proceedings, Cincinnati, 21-23 May. Technology Park, Atlanta, Georgia, TAPPI, pp.139-51

Anon., 1976. Paper and paperboard manufacturing. (39-40). Coating of paper and board. Paper, 186(8):484,487,489-90,496; 186(9):561-2,565-6,569

_____, 1980. Alginate fibres create textures of fruits, vegetables and meats. Food Process.,Chicago, 41(7):22-4

_____, 1983. New concept for growing market: structured potatoes. Food Eng., 55(5):72-3

_____, 1983a. Gums - jellies - pastilles. 6. Alginates: incorporating fruit pulps: Turkish delight. Confect. Prod., 49(11):586-8

CHAPTER 3

PRODUCTION, PROPERTIES AND USES OF CARRAGEENAN

by

Norman Stanley
FMC Corporation, Marine Colloids Division
5 Maple Street, Rockland
Maine 04841, USA

CARRAGEENANS

INTRODUCTION

Carrageenans are commercially important hydrophilic colloids (water-soluble gums) which occur as matrix material in numerous species of red seaweeds (Rhodophyta) wherein they serve a structural function analogous to that of cellulose in land plants. Chemically they are highly sulfated galactans. Due to their half-ester sulfate moieties they are strongly anionic polymers. In this respect they differ from agars and alginates, the other two classes of commercially exploited seaweed hydrocolloids. Agars, though also galactans, have little half-ester sulfate and may be considered to be nonionic for most practical purposes. Alginates, though anionic, are polymers of mannuronic and guluronic acids and as such owe their ionic character to carboxyl rather than sulfate groups. In this respect alginates are more akin to pectins, found in land plants, than to the other seaweed hydrocolloids.

Applications of carrageenans make use of both their hydrophilic and anionic properties, the latter influencing the former.

SOURCES

In the West the algae _Chondrus crispus_ and _Gigartina stellata_, as the sun-bleached whole plants, have been used for centuries for making jellies and milk puddings (blanc mange). An old recipe for blanc mange (Smith, 1905) is as follows:

"Soak half a cup of dry moss in cold water for five minutes, tie in a cheesecloth bag, place in a double boiler with a quart of milk and cook for half an hour; add half a teaspoonful of salt or less, according to taste, strain, flavor with a teaspoonful of lemon or vanilla extract as desired, and pour into a mold or small cups, which have been wet with cold water; after hardening, eat with sugar and cream."

"Moss" here refers to Irish moss, a common name for Chondrus crispus. Irish moss has also been known as carrageen from the Irish word, carraigeen, meaning "rock moss."

Stanford (1862) coined the name "carrageenin" for the gelatinous material extracted by water from Chondrus crispus. The present spelling, "carrageenan", has become accepted within the past 25 years, this being consonant with the use of the -an suffix for the names of polysaccharides.

In the Far East marine algae of the genus Eucheuma in the Solieriaceae family have a long history of use as articles of food, for their supposed medicinal properties, and in trades such as bookbinding wherein their mucilage is used as an adhesive (Eisses 1953; Zaneveld 1955, 1959). Although the Malayan word, "agar-agar" refers to Eucheuma species it is now known that these yield carrageenans rather than agar-type polysaccharides.

Due to the importance of sea plants in the economy of Indonesia and the general area of the Malay peninsular sporadic attempts were made to set up programs for the comprehensive investigation of the marine algae of the area and the products obtainable from them. Unfortunately, due to the underdeveloped economies and unsettled political conditions in these countries, none of the proposed programs was ever carried out to the extent comtemplated. The principal results of these investigations have been reported by Eisses (1952, 1953) and Zaneveld (1955, 1959). They describe a number of types of Eucheuma which have been harvested in this area, listing them by their botanical and native names. It was not until the Eucheumas were recognized as valuable carrageenophytes by the western carrageenan industry that the large-scale export of these species became established. The introduction of Eucheuma farming in the Philippines in 1971 greatly promoted this industry.

From the original identification of Chrondrus crispus as a carrageenophyte this classification is now known to cover numerous species from seven different families: Solieriaceae, Rhabdoniaceae, Hypneaceae, Phyllophoraceae, Gigartinaceae, Furcellariaceae, and most recently, Rhodophyllidaceae (Deslandes et al., 1985). While not all of these have been, or are perhaps likely to be, exploited commercially, present-day sources of carrageenans go well beyond the original Irish moss. Seaweeds which have been used for carrageenan production include Chondrus crispus, C. ocellatus, Gigartina stellata, G. acicularis, G. pistillata, G. canaliculata, G. chamissoi, G. radula (also identified in the literature as Iridea species), G. skottsbergii, Gymnogongrus furcellatus, Eucheuma cottonii, E. spinosum, E. gelatinae, Furcellaria fastigiata, Hypnea musciformis, and H. spicifera. Utilization of these not only has greatly extended the base, and the geographical area, from which the industry can draw raw materials, but also has extended the range of properties of their extractives, as different species yield carrageenans of differing structure and properties.

Chondrus crispus is largely harvested in the Maritime Provinces of Canada with smaller quantities collected along the coasts of Maine and Massachusetts in the United States. The difference in volume is due not so much to lack of abundance as to differences in the local economies. The carrageenan from C. crispus, which comprises a mixture of kappa- and lambda-carrageenan, is much valued as the preferred type for applications such as chocolate milk stabilization.

It is now known that kappa- and lambda-carrageenan do not occur together in the same plant, but are elaborated at different stages of the reproductive cycle. Kappa-carrageenan occurs in the haploid gametophytic plants and lambda in the diploid tetrasporophytes (Chen et al., 1973). Since these occur and are harvested together a mixture of kappa and lambda is obtained on processing. This is true not only of Chondrus but of other species of Gigartinaceae and Phyllophoraceae as well (McCandless, West and Guiry, 1982, 1983). It has not been found to hold true for carrageenophytes in the Solieriaceae family (Doty and Santos, 1978) or, as far as is known, in those of other families.

Gigartina acicularis and G. pistillata occur and are harvested together along the coasts of southern France, Spain, Portugal, and Morocco. The latter two species are unique in that they yield a nongelling, predominantly lambda or xi type carrageenan.

Gigartina radula is harvested in Chile and comprises a major resource for carrageenan production. The taxonomy of Chilean G. radula is a subject of controversy, with some algologists holding to the earlier classification of this seaweed as one or more Iridea species.

Eucheuma cottonii and E. spinosum are now heavily used by carrageenan producers and are harvested in large quantities in Indonesia and the Philippines. Farming of these Eucheuma species is now practiced on a large scale in the Philippines, and this has done much to increase and stabilize the supply of these important carrageenophytes. E. cottonii and E. spinosum are remarkable in that these two species, once thought to be varieties of a single species, yield quite different types of carrageenans. E. cottonii, which actually may comprise two very similar species, E. cottonii and E. striatum (Doty, 1973), yields nearly ideal kappa-carrageenan, while E. spinosum yields nearly ideal iota-carrageenan.

Gymnogongrus furcellatus, harvested in Peru, has been used as a source of iota-type carrageenan. Supplies of this weed, however, are small.

Furcellaria fastigiata yields furcellaran ("Danish agar"), often treated in the literature as a polysaccharide distinct from carrageenan but now considered, on the basis of chemical evidence (Lawson, et al., 1973), to be a member of the carrageenan family of polysaccharides. Although F. fastigiata is found along many coasts of the North Atlantic and its adjacent seas major quantities are harvested only in Denmark and the Maritime Provinces of Canada. It grows both in an attached and unattached form, the latter reproducing only vegetatively. A large body of unattached Furcellaria in the Kattegat, near the coast of Jutland, was for many years the sole source of raw material for furcellaran production (Lund and Bjerre-Petersen, 1952, 1961). In some Canadian locations Furcellaria grows alone, but most commonly it grows in mixed beds with Chondrus crispus. It is collected principally on the coasts of Prince Edward Island and Nova Scotia after being cast up on the beaches. The unattached form of Furcellaria has not been found in Canada (Bjerre-Petersen, Christensen and Hemmingsen, 1973).

Hypnea musciformis has been harvested along the south-eastern coast of the United States, in Brazil, and in Senegal. It yields a kappa or furcellaran type carrageenan. It is no longer used by the major carrageenan manufacturers due to difficulty in processing and low yield.

Table 1 lists annual world production of carrageenophytes by producing area over the past 15 years. The large increase over this period is principally due to the growth of Eucheuma farming in the Philippines. In most other areas production has remained relatively static.

Table 1

Production of carrageenan seaweed
world basis in metric tons

Countries	1971	1979	1984
Canada	6 000	5 700	5 000
Philippines	500	14 000	25 000
Chile	4 000	6 000	6 000
Indonesia	4 000	3 500	3 000
Others	5 500	4 500	4 500
Total	20 000	33 700	43 500

CHEMICAL COMPOSITION

Carrageenans have the common feature of being linear polysaccharides with a repeating structure of alternating 1,3-linked β-D-galactophyranosyl and 1,4-linked α-D-galactopyranosyl units (Figure 1). The 3-linked units occur as the 2- and 4-sulfate, or unsulfated, while the 4-linked units occur as the 2-sulfate, the 2,6-disulfate, the 3,6-anhydride, and the 3,6-anhydride-2-sulfate. Sulfation at C3 apparently never occurs. Pyruvate has been reported present in the carrageenans from some Gigartina species; these carrageenans have been termed "pi-carrageenan" (Hirase and Watanabe, 1972; DiNinno, McCandless and Bell, 1979; McCandless and Gretz, 1984). Methoxyl groups occur in sulfated polysaccharides from the Grateloupiaceae family (Hirase, Araki and Watanabe, 1967; Nunn and Parolis, 1968; Allsobrook, Nunn and Parolis, 1971). There is some question, though, as to whether these have the alternating structure characteristic of carrageenans (Parolis, 1981).

In their original work on fractionation of carrageenan from Chondrus crispus with potassium chloride Smith and Cook (1953) isolated two fractions which they named kappa- and lambda-carrageenan. Kappa was defined as that fraction which was precipitated by potassium chloride, while lambda was the fraction which remained in solution. Chemical studies on these fractions revealed that nearly half of the sugar units in kappa were 3,6-anhydro-D-galactose, a sugar not previously known to occur in nature, while lambda contained little or none of this sugar (Smith, Cook and Neal, 1954).

Due largely to the investigations by Rees and his co-workers (Rees 1963; Dolan and Rees, 1965; Anderson, Dolan and Rees, 1968, 1973; Anderson et al., 1968, 1968a; Lawson and Rees, 1968; Lawson et al., 1973; Penman and Rees, 1973, 1973a, 1973b;) carrageenans are now defined in terms of chemical structure. While it is true that more or less of a continuous spectrum of carrageenans exists (Pernas et al., 1967), it is nevertheless possible to distinguish a small number of ideal or limit polysaccharides. The names mu, kappa, nu, oita, lambda, theta, and xi are presently applied to these limit carrageenans. Figure 2 shows the repeating units of these polysaccharides. Mu and nu are believed to be precursors in the biosynthesis of kappa and iota respectively (Anderson et al., 1968b; Stancioff and Stanley, 1969), the transformation's being accomplished in the alga by an enzyme, "dekinkaase" (Lawson and Rees, 1970), or, in industrial processing, by the base-catalyzed S_n2 elimination of 6-sulfate (Stanley, 1963). Lambda likewise can be at least partially converted to theta-carrageenan by this reaction, but theta has yet to be identified as occurring naturally.

Xi-carrageenan, which constitutes the KCl-soluble fraction of some Gigartina species (G. chamissoi and G. canaliculata), has not been completely characterized but seems to differ from lambda in that

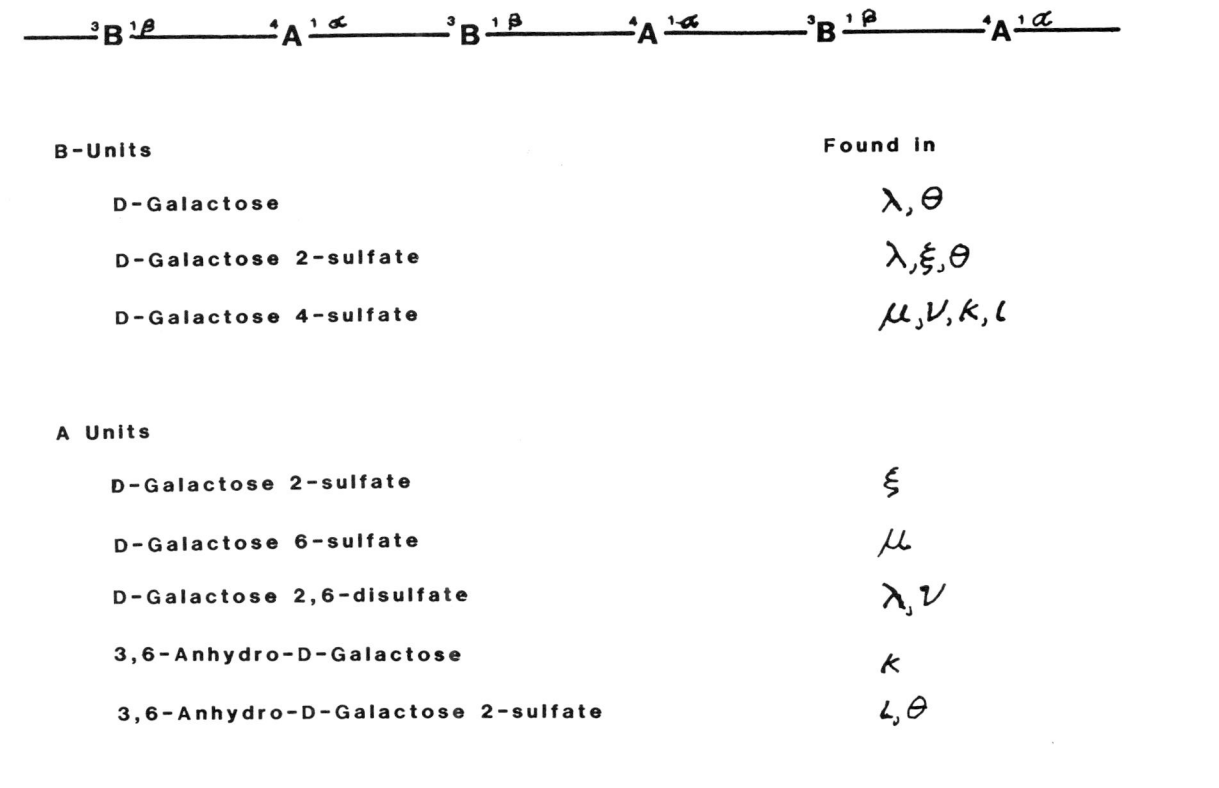

$$\underline{\quad}^3B^{1\beta}\underline{\quad}^4A^{1\alpha}\underline{\quad}^3B^{1\beta}\underline{\quad}^4A^{1\alpha}\underline{\quad}^3B^{1\beta}\underline{\quad}^4A^{1\alpha}$$

B-Units	Found in
D-Galactose	λ, θ
D-Galactose 2-sulfate	λ, ξ, θ
D-Galactose 4-sulfate	μ, ν, κ, ι

A Units	
D-Galactose 2-sulfate	ξ
D-Galactose 6-sulfate	μ
D-Galactose 2,6-disulfate	λ, ν
3,6-Anhydro-D-Galactose	κ
3,6-Anhydro-D-Galactose 2-sulfate	ι, θ

Figure 1 Repeating structure of carrageenans (Reproduced with permission from Handbook of water-soluble gums, edited by R.L. Davidson, New York, McGraw-Hill, 1980)

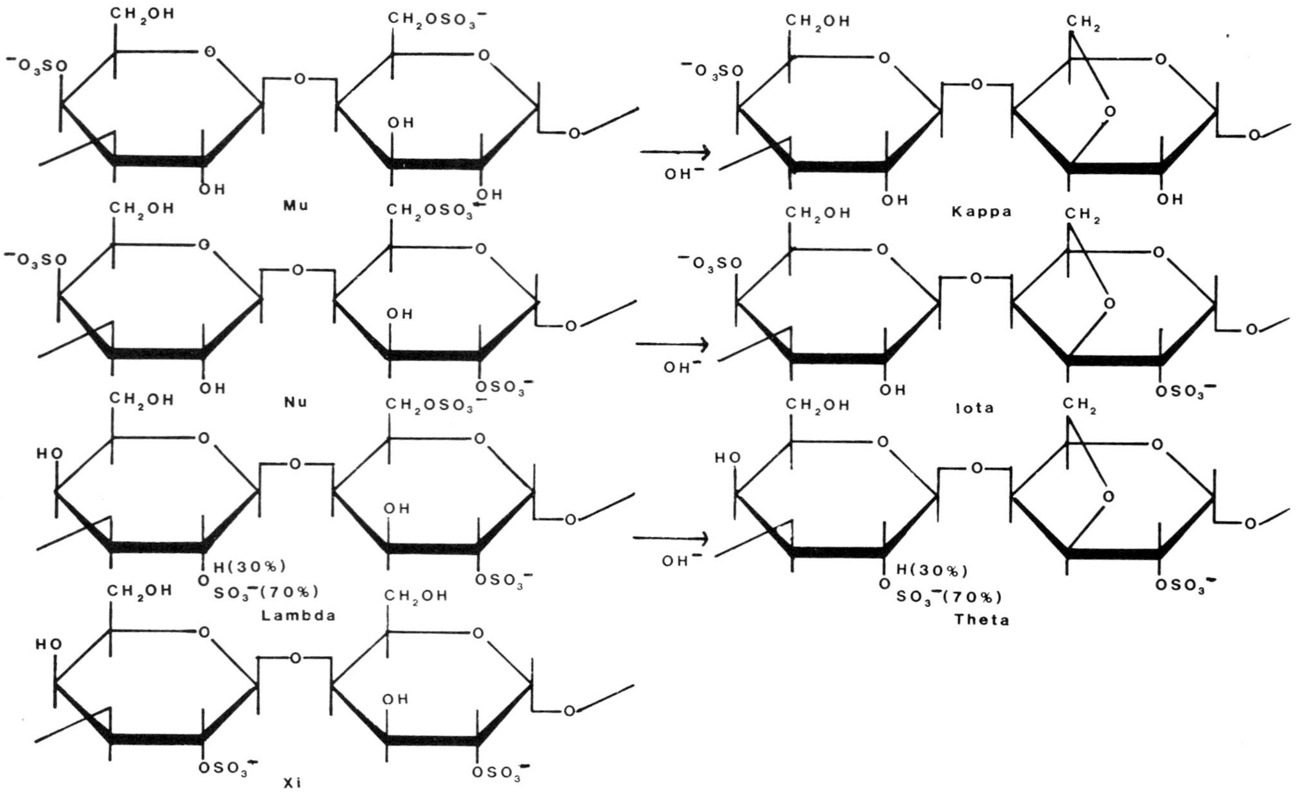

Figure 2 Repeating units of limit carrageenans (Reproduced with permission from Handbook of water-soluble gums, edited by R.L. Davidson. New York, McGraw-Hill, 1980)

the 1,3-linked units are unsubstituted at C6 (Penman and Rees, 1973a). Sugar units lacking sulfate at this position cannot be converted to the anhydride.

A new family of carrageenans for which the 3-linked units are unsulfated has recently been reported as the polysaccharide of Eucheuma gelatinae. This family consists of beta-carrageenan, analogous to kappa but lacking sulfate on C4 of the 1,3-linked units, and its precursor, gamma-carrageenan, analogous to mu (Greer and Yaphe, 1984).

Furcellaran, the polysaccharide from Furcellaria fastigiata, is very much like kappa-carrageenan, differing mainly in the amount of half-ester sulfate present. Furcellaran contains one sulfate group per three or four sugar units, as compared with one sulfate group per two sugar units for kappa-carrageenan. D-galactose 2-sulfate, D-galactose 4-sulfate, D-galactose 6-sulfate, and 3,6-anhydro-D-galactose 2-sulfate have been identified as components of furcellaran (Painter, 1966). The distribution of sulfate along the molecular chain is still not completely known.

Native furcellaran, like carrageenan, can be modified by treatment with hot alkali to cleave 6-sulfate, with formation of 3,6-anhydro-D-galactose units. As with kappa-carrageenan this results in an increase in water and milk gel strengths.

Precipitation of furcellaran with potassium chloride occurs at much lower concentrations than does precipitation of carrageenans (Smidsrød et al., 1967). This is due to furcellaran's having fewer charged groups and therefore being less hydrophilic than other carrageenans.

Native carrageenans from different algae may be regarded as varying mixtures of the limit polysaccharides and intermediate hybrids ranging in degree of anhydridation and 2-sulfation of the 1,4-linked units. This is shown graphically in Figure 3. This diagram divides the carrageenans into two general groups. One consists of mu, nu, kappa, iota, and their hybrids. Carrageenans in this group gel with potassium ions, or can be made to gel by treatment with alkali; they are characterized by having their 1,3-linked units either unsulfated or sulfated only at C4. The other group consists of lambda, xi, theta, and their hybrids; these do not gel either before or after alkali treatment and characteristically have both their 1,4- and 1,3-linked units sulfated at C2, though occasionally the latter are unsulfated (Stancioff and Stanley, 1969).

The complex fine structure of carrageenans is still an active field of research. Enzymic, immunological, and ^{13}C NMR techniques have proved to be powerful tools for these investigations.

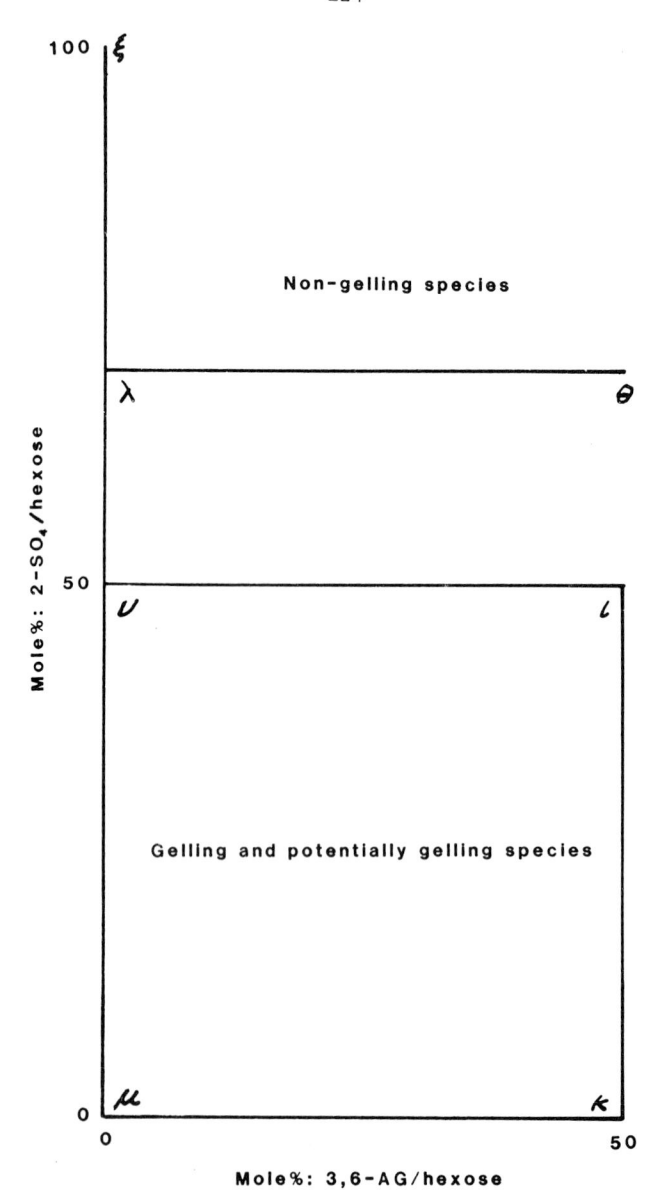

Figure 3 Composition of carrageenans. (Reproduced with permission from Handbook of water-soluble gums, edited by R.L. Davidson. New York, McGraw-Hill, 1980)

EXTRACTION PROCESSES

Specific details of extraction processes are closely guarded as trade secrets by the several manufacturers of carrageenans, but broadly these follow a similar pattern. Weed, usually dried and baled, is received at the processing location from the harvesting location. The shipment may be sampled and the sample subjected to a test extraction to evaluate the quality of the extractive. Other eed quality factors such as contents of moisture, sand and salt, and non-carrageenophytes are evaluated at this stage. Obtaining a representative sample from a weed shipment is not a trivial exercise, as weed quality may vary widely not only from one shipment to the next but also within a shipment, due to factors over which the processor may have but limited control. Sampling protocols used generally represent what is feasible rather than what a statistician might regard as adequate.

Prior to plant-scale extraction the weed may be washed to remove adhering salts, sand, stones, and marine organisms. Washed, or unwashed, weed, usually as a blend selected to achieve the desired properties in the extractive, is then digested with hot water under alkaline conditions to exhaustively extract the carrageenan. The alkali, usually calcium or sodium hydroxide, performs two functions: firstly it promotes swelling and maceration of the weed to aid in bringing the carrageenan into solution, while, secondly, when employed at sufficiently high concentrations, it effects cleavage of 6-sulfate groups from the carrageenan to generate 3,6-anhydro-D-galactose residues in the polysaccharide chain. These function to enhance the water gel strength and milk reactivity of the carrageenan. Maceration is promoted by agitation of the resultant paste. Conversion of 6-sulfated moieties to the 3,6-anhydride continues during resting of the paste at temperatures near $100^{\circ}C$.

When the desired degree of conversion has been achieved the solution of carrageenan is separated from weed solids by filtration, or by centrifugation followed by filtration. Concentration of the filtrate by evaporation, and adjustment of pH, are done prior to the recovery of the carrageenan from solution.

The foregoing processing operations inevitably involve some degradation of the polysaccharide, due to the rigors (e.g., heat, alkalinity) of processing. Although carrageenans are reasonably stable under the conditions of alkalinity encountered in processing a drop in pH can occur from the consumption of alkali for the neutralization of sulfuric acid formed by cleavage of half-ester sulfate groups. Saccharinic acids may also be formed through alkali-catalyzed "peeling" reactions.

Several methods have been used to recover the carrageenan from solution. Direct drying of the concentrated filtrate on steam-heated rolls has been used extensively. Products of much higher quality are obtained by precipitation of the carrageenan from solution by 2-propanol or other alcohols. An interesting historical note is that perhaps the earliest process described for recovering carrageenan from Irish moss employed alcohol precipitation (Bourgade, 1871).

Precipitation is followed by further alcohol washes to dehydrate the coagulum. Vibrating screens or basket centrifuges may be used to separate the coagulum from the alcohol following precipitation and each wash. Following the final wash the coagulum is dried under conditions permitting recovery of the residual alcohol.

The fibrous carrageenan from the dryer is ground and sifted to specified particle sizes which may range from 80 mesh to 270 mesh. This basic product, segregated into batches, is sampled and tested for compositional and functional properties (e.g., moisture, viscosity, gel strength).

Another process presently in use for the recovery of carrageenan from solution was originally developed for furcellaran production but is also employed for kappa-carrageenan. This takes advantage of several properties common to furcellaran and kappa-carrageenan. First, solutions of these polysaccharides form gels in the presence of potassium ions. Second, these gels exude water by syneresis on standing, the more so when squeezed in a press. Third, much water separates from the gel when it is frozen and then allowed to thaw. The latter phenomenon is the same as that used for the production of agar.

In the case of Furcellaria, the weed may be treated in the cold with an alkaline solution for one or more weeks. This alkaline treatment removes colouring matter and some proteins and makes the gum more easily extractable. Some alkaline elimination of 6-sulfate may also occur during this treatment. Extraction follows a procedure generally similar to that described above for the alcohol process. Following concentration by evaporation the filtrate is extruded through spinnerets into a cold 1-1.5% solution of potassium chloride. The resulting gelled threads are further dewatered by subsequent potassium chloride washes followed by pressing. The gel is then frozen, thawed, chopped, again washed with fresh KCl solution, and air-dried.

A limitation of the freeze-thaw process as applied to carrageenans is that it is applicable only to furcellaran and kappa-carrageenan, which are the only types whose gels with potassium ions exhibit marked syneresis. Moreover the requirement that potassium be present precludes making products wherein sodium is the major counterion.

The "gel-press" process, used by some minor producers of carrageenans and agar, likewise relies on pressure to dewater the gel, but omits the freeze-thaw cycle.

The economics of extraction processes are strongly affected by the cost of the energy required to bring the carrageenan into solution and subsequently to recover it in dry form. This includes the heat necessary to digest and cook the seaweed, to concentrate the filtrate in evaporators, to dry the coagulum, and, in the case of alcohol precipitation, to recover the spent alcohol by distillation. Steam, generated by oil-fired boilers, is the usual source of process heat. Given the volatility of oil prices in the present-day market it will be appreciated that the cost of energy from this source has changed drastically in the past and can be expected to do so in the future. The impact of energy cost can of course be less in any locality where local low-cost fuels can be exploited. As an example, one major manufacturer of carrageenans uses locally-mined peat as a source of process heat.

High energy costs can be countered by employing cogeneration to supply the not inconsiderable electrical power requirements for plant operation. Alternatives to evaporation (e.g., ultrafiltration) have been investigated. These capital-intensive measures are not presently economical, but may be expected to become so in the likely event of rising energy costs.

An adequate supply of cheap, good quality, fresh water is an obvious prerequisite for the economical operation of an extraction process. In at least one instance a carrageenan factory was forced to relocate to another area when it was found that its expanding demands exceeded the capacity of the local water supply.

Filtration must be employed if refined, completely water-soluble, products are to be produced. Owing to the viscosity of the extract and the swollen, gelatinous nature of the residual solids pressure filters are a necessity for efficient throughput and a filter aid must be added to the feed to prevent clogging ("blinding") of the filter medium and to afford a porous filter cake that will drain well and can be washed in the filter press to recover retained carrageenan. The filter aids most commonly employed are calcined diatomaceous earth and expanded volcanic glass. Since large amounts of aid are required the choice may depend on the location of the carrageenan processing plant with respect to the source of supply of aid.

Diatomaceous filter aids are available in a range of grades of retentiveness. A retentive aid may be used for a secondary "polish" filtration of the effluent from the primary filtration. This may be preceded by treatment of the primary filtrate with activated carbon to decolorize it. However this practice is now uncommon as the expense of these added steps usually cannot be recouped as added value of the product. Moreover the tendency of carbon to peptize and pass through even the most retentive filter can result in the product's having an undesirable grayish color.

It has been estimated that for a carrageenan extraction plant to be economical it should have a capacity of at least 450-750 metric tons of product annually. This would require processing 1 400-2 300 MT of dried seaweed. To allow for variations in the harvest, available seaweed sources should exceed these tonnages by 40-50%. Initial investment for procurement, production, and marketing has been estimated to be $ 4 500 000 - $ 6 750 000. While these may be taken as ballpark estimates, obviously many factors other than scale of production determine whether a given operation will prove profitable. It has been recommended that any move toward production should be preceded by a three- to four-year pilot program of harvesting and selling dried seaweeds to other producers. This will serve to develop information required for starting an extraction plant (Moss, 1978).

A new process, and product, is semi-refined carrageenan. This process, which is distinguished by its low evergy input, uses Eucheuma cottonii as a raw material.

Semi-processed E. cottonii is prepared by a method which superficially resembles that for French fried potatoes. A basket of seaweed fronds is immersed and cooked in hot aqueous potassium hydroxide and then soaked in fresh water to extract most of the residual alkali. The product is dried and ground to produce a flour having many of the properties of conventional extracted carrageenans.

The economic advantage lies in not extracting the carrageenan from the seaweed but rather performing the reaction which maximizes gel strength upon the polymer within the plant structure. By doing this the ratio of process water to product is minimized, thereby reducing the cost of isolating the dry product.

The rationale for the process is as follows: the kappa-carrageenan in E. cottonii does not dissolve in hot water if the concentration of gel-inducing cations (e.g., K^+, Ca^{++}) is maintained at a level corresponding to a melting temperature for the in situ carrageenan gel which is greater than the processing temperature. If this is done the carrageenan will imbibe water to form a gel structure like that in the living plant. The reagent (OH^- ions) is able to diffuse into this structure and produce the chemical modification desired.

PROPERTIES

The chemical reactivity of carrageenans is primarily due to their half-ester sulfate groups which are strongly anionic, being comparable to inorganic sulfate in this respect. The free acid is unstable, and commercial carrageenans are available as stable sodium potassium and calcium salts or, most commonly, as a mixture of these. The associated cations together with the conformation of the sugar units in the polymer chain determine the physical properties of the carrageenans. For example, kappa- and iota-carrageenans form gels in the presence of potassium or calcium ions whereas lambda-carrageenan does not.

Reactivity with proteins is exhibited by both gelling and nongelling carrageenans, although regularity of the polymer is an important factor. In most, if not all, cases ion-ion interactions between the sulfate groups of the carrageenan and the charged groups of the protein are involved. Reaction depends on protein/carrageenan net charge ratio, and thus is a function of the isoelectric point of the protein, the pH of the system, and the weight ratio of carrageenan to protein (MacMullan and Eirich, 1963). At pH levels below the isoelectric point the protein has a net positive charge and thus can undergo direct electrostatic interaction with the negatively-charged carrageenan. The commercially important reaction of carrageenan with kappa-casein in milk is specific for this protein and unique in that it can occur at pH levels above the isoelectric point of the casein. It appears to be due to a region of positively charged amino acid residues in the kappa-casein molecule which can interact electrostatically with the sulfate groups of the carrageenan even though the net charge of the casein is negative (Snoeren et al., 1975).

The functionality of carrageenans in various applications depends largely on their rheological properties. Carrageenans, as linear, water-soluble, polymers, typically form highly viscous aqueous solutions. This is due to their unbranched, linear macromolecular structure and polyelectrolytic nature. The mutual repulsion of the many negatively charged half-ester sulfate groups along the polymer chain causes the molecule to be highly extended, while their hydrophilic nature causes it to be surrounded by a sheath of immobilized water molecules. Both of these factors contribute to resistance to flow.

Viscosity depends on concentration, temperature, the presence of other solutes, and the type of carrageenan and its molecular weight. Viscosity increases nearly exponentially with concentration. This behaviour is typical of linear polymers carrying charged groups and is a result of the increase with concentration of interaction between polymer chains. Salts lower the viscosity of carrageenan solutions by reducing the electrostatic repulsion among the sulfate groups. This behaviour is likewise normal for ionic macromolecules. At low temperature and high enough salt concentration, however, carrageenan solutions may gel, with an increase in apparent viscosity. This is particularly true for the strongly gel-inducing cations, K^+ and Ca^{++}.

At high temperatures, however, Ca^{++} lowers viscosity to a greater extent than does Na^+ or K^+.

Viscosity decreases with temperature. Again, the change is exponential. It is reversible provided that heating is done at or near the stability optimum at pH 9, and is not prolonged to the point where significant thermal degradation occurs. Both gelling (kappa-, iota-) and nongelling (lambda-) carrageenans behave in this manner at temperatures above the gelling point of the carrageenan. On cooling, however, the gelling types will abruptly increase in apparent viscosity when the gelling point is reached, provided that the counter-ions (K^+ and Ca^{++}) promotive of gelation are present.

Viscosity increases with molecular weight in accordance with the Mark-Houwink equation:

$$[\eta] = KM^\alpha$$

where $[\eta]$ is intrinsic viscosity, M is an average molecular weight (since carrageenans are polydisperse) and K and α are constants. Intrinsic viscosities correlate well with practical viscosity measurements taken at 1.5% concentration and 75°C.

Commercial carrageenans are available in viscosities ranging from about 5 mPa.s to 800 mPa.s when measured at 1.5% concentration and 75°C. Carrageenan solutions having viscosities less than 100 mPa.s have flow properties very close to Newtonian. At higher viscosities the solutions exhibit shear-thinning behaviour and it becomes necessary to specify the shear rate at which the measurement is taken. Where non-Newtonian behaviour is expected viscosity measurements should be made at a shear rate comparable to that encountered in the application considered.

Carrageenans specifically tailored for water-thickening applications are usually lambda types or the sodium salt of mixed lambda and kappa. They dissolve in either cold or hot water to form viscous solutions. There high water viscosities are desirable, and the high molecular weight and hydrophilicity of lambda contribute to this. For gelling applications a low viscosity in hot solution is usually desirable for ease in handling, and, fortunately, high gel strength carrageenans (mixed calcium and potassium salts of kappa or iota) fulfill this requirement because of their lesser hydrophilicity and the effect of the calcium ions.

Kappa- and iota-carrageenans and furcellaran form gels on cooling of their hot solutions in the presence of certain cations, notably K^+ and Ca^{++}. Heating is required to bring them into solution under these conditions. According to Rees (1972) carrageenans which form aqueous gels do so because of double helix formation. At temperatures above the melting point of the gel thermal agitation overcomes the tendency to form helices and the polymer exists in solution as random coils. On cooling the polymer chains become interlinked through double helix formation to form "domains" (Morris, Rees and Robinson, 1980). This occurs regardless of the counterions present and does not directly lead to gelation. Only when potassium or other gel-promoting cations are present will the domains aggregate to form a three-dimensional network. An alternative model of carrageenan gelation, based on cation-induced aggregation of single helices has also been proposed (Paoletti, Smidsrød and Grasdalen, 1984).

Regardless of the mechanism it appears that the occurrence of 1,4-linked 6-sulfated residues in the polymer chain of either kappa- or iota-carrageenan detracts from the strength of their gels. This is ascribed to kinks, produced by these residues, in the chain which inhibit the formation of double helices (Mueller and Rees, 1968). Alkali modification of the carrageenan during processing increases the gel strength of the product by removal of these kinks through conversion of 6-sulfated residues to the 3,6-anhydride. Increased hydrophobicity from the added anhydride residues may also contribute to gelation.

Kappa-carrageenan and furcellaran gels are relatively rigid and are subject to syneresis. Incorporation of locust bean galactomannan along with kappa or furcellaran yields a more compliant gel. "Smooth" regions of the mannan chain (i.e., regions carrying no galactose side groups) are believed to bind to the double helices of the kappa to reduce their tendency to aggregate (Rees, 1972). Iota-carrageenan by itself yields compliant gels with very little tendency to undergo syneresis. Here the 2-sulfate groups on the 3,6-anhydride residues act as wedging groups to prevent the tightly-packed aggregation believed responsible for the rigidity of kappa gels. Whereas potassium is more effective than calcium in inducing gelation of kappa the reverse is true for iota-carrageenan.

All carrageenans have the ability to form gels by cooling a solution of the carrageenan in hot milk. Even lambda-carrageenan, which does not gel in water regardless of the cations present, will form a gel at levels of 0.2% or more by weight of the milk. This gelation is ascribed to the formation of carrageenan-casein bonds, as previous described.

With kappa- and iota-carrageenan as well as furcellaran there is, in addition to the carrageenan-casein interaction, a water-gelling effect from the cations present in the carrageenan as well as the Ca^{++} and K^+ present in the milk. These cations appear to be required for milk gel formation as milk which has been ion-exchanged to remove Ca^{++} and K^+ does not gel with the sodium salts of lambda, kappa, or iota. On the other hand the strength of milk gels is enhanced by the addition of soluble calcium and potassium salts in a manner quite similar to that in water gels.

The presence of fat influences the behaviour of carrageenans in milk. Strongly gelling kappa-carrageenans can be used in high-fat systems but are not well tolerated in low-fat systems wherein they may exert a destabilizing action resulting in whey separation. For the latter, weak milk-gelling kappa-carrageenans with high ester sulfate and moderate to high 3,6-anhydride are more suitable. The reason that strongly-gelling types can be employed in high-fat but not low-fat systems is due in part to the disperse fat phase. This apparently tempers the carrageenan-casein complex, serving to interrupt aggregation to some extent. Interaction may also occur between carrageenan and the phospholipid which is present as a monomolecular layer covering the disperse globules of butterfat in the milk. Since the phospholipid contains basic amino groups with which the ester sulfate groups of the carrageenan can react it is very likely that electrostatic bonds are formed between the carrageenan and phospholipid. This may account for the extraordinary effectiveness with which very low levels (ca. 50 ppm) of carrageenan stabilize evaporated milk against fat separation (Moirano, 1977).

Carrageenans are susceptible to depolymerization through acid-catalyzed hydrolysis. At high temperatures and low pH this may rapidly lead to complete loss of functionality. They can be used in acid systems, however, if not subjected to prolonged heating. The rate of hydrolysis at a given pH and temperature is markedly lower if the carrageenan is in the gel rather than the sol state. This can be achieved by ensuring that gel-promoting cations are present at sufficient concentration to raise the gel melting temperature above the temperature to which the carrageenan will be subjected.

Carrageenan is listed by the U.S. Food and Drug Administration (FDA) as Generally Recognized as Safe (GRAS) (U.S. Food and Drug Administration, 1979). Following reports of cecal and colonic ulceration in guinea pigs and rabbits induced by a highly degraded carrageenan provided, ironically, for the symptomatic relief and cure of peptic and duodenal ulcers in man, intensive investigations into the safety of carrageenans were carried out by the FDA and other groups sponsored by the carrageenan industry. By late 1976 food grade carrageenan, defined as having a water viscosity of no less than 5 mPa.s at 1.5% concentration and 75°C (U.S. Food and Nutrition Board, 1981) had been demonstrated to be safe.

Carrageenans have been shown not to be teratogenic (Collins, Black and Prew, 1977, 1977a, 1979). A study conducted on rats and hamsters at the Eppley Institute for Cancer Research demonstrated that carrageenans are not carcinogenic. A review of the physiological effects of carrageenans has been published (Stancioff and Renn, 1975).

CURRENT USES

Carrageenans are used to gel, thicken, or suspend; therefore they are used in emulsion stabilization, for syneresis control, and for bodying, binding and dispersion. Major uses are in foods, particularly dairy applications. Tables 2 and 3 list dairy and water-based applications respectively (Guiseley, Stanley and Whitehouse, 1980).

Furcellaran generally finds applications similar to those for kappa-carrageenan. Historically, furcellaran has dominated two major European application fields: Tarte or cake glaze powders and flan powders. Today the special properties of excellent gel texture and flavour release make furcellaran a preferred product for use in milk pudding powders.

Carrageenan is unique in its ability at very low concentration (ca. 300 ppm) to suspend cocoa in chocolate milk; no other gum has been found to match it. A very delicate milk gel structure, undetectable on pouring or drinking the milk, is believed to hold the cocoa in suspension. A substantial differential ("spread") between the concentration at which settling of cocoa occurs and that at which visible gelation is evident is required for practical stabilization. This is achieved by careful selection of weed type and quality.

The use of iota-carrageenan in dessert gel formulations affords gels which have textures very similar to those of gelatin gels. They have an advantage over gelatin gels in that their melting point is higher, so that they find a ready market in tropical climates or where refrigeration is not available. This is offset to some extent by the different mouth-feel, since these gels do not "melt in the mouth", as does gelatin. A further advantage is that iota gels retain their tender structure on aging, whereas gelatin tends to toughen. This is important for ready-to-eat desserts, an item popular in Europe.

Kappa-carrageenan or furcellaran by itself is unsatisfactory for dessert gel applications due to the "short", brittle structure of its gel. This can be ameliorated by the incorporation of locust bean galactomannan into the formulation, and kappa-locust bean or iota-kappa-locust bean blends are also offered for this application. To achieve sparkling-clear gels it is necessary to use a locust bean gum which has been clarified by filtration. The clarified gum is produced for this purpose by several of the major carrageenan manufacturers.

In toothpastes carrageenans function as a "binder" to impart the desired rheological properties to the paste and to provide the cosmetic quality of "sheen". Toothpastes consist of ingredients which interact in complex and poorly understood ways and the carrageenan often must be carefully tailored to achieve satisfactory performance in a particular formulation. Carrageenan suffers severe competition in the U.S. domestic market from sodium carboxymethylcellulose, a much cheaper gum. Despite this, business has been retained - and regained - due to the superior quality and appearance carrageenan imparts to a toothpaste. Outside the United States carrageenan has maintained a

Table 2

Typical milk (dairy) applications of carrageenan

Use	Function	Product	Approx. use level, %
Frozen desserts:			
Ice cream, ice milk	Whey prevention		
	Control meltdown	Kappa-	0.010-0.030
Pasteurized milk products:			
Chocolate, eggnog, fruit-flavored	Suspension, bodying	Kappa-	0.025-0.035
Fluid skim milk	Bodying	Kappa-, iota-,	0.025-0.035
Filled milk	Emulsion stabilization, bodying	Kappa-, iota-	0.025-0.035
Creaming mixture for cottage cheese	Cling	Kappa-	0.020-0.035
Sterilized milk products:			
Chocolate, etc.	Suspension, bodying	Kappa-	0.010-0.035
Controlled calorie	Suspension, bodying	Kappa-	0.010-0.035
Evaporated	Emulsion stabilization	Kappa-	0.005-0.015
Infant formulations	Fat and protein stabilization	Kappa-	0.020-0.040
Milk gels:			
Cooked flans or custards	Gelation	Kappa-, Kappa- + iota-	0.20-0.30
Cold prepared custards			

Table 2 (continued)

Use	Function	Product	Approx. use level,%
(with added TSPP)	Thickening, gelation	Kappa-, iota-, lambda-	0.20-0.50
Pudding and pie fillings (starch base)			
Dry mix cooked with milk	Level starch gelatinization	Kappa-	0.10-0.20
Ready-to-eat	Syneresis control, bodying	Iota-	0.10-0.20
Whipped products:			
Whipped cream	Stabilize overrun	Lambda-	0.05-0.15
Aerosol whipped cream	Stabilize overrun stabilize emulsion	Kappa-	0.02-0.05
Cold prepared milks:			
Instant breakfast	Suspension, bodying	Lambda-	0.10-0.20
Shakes	Suspension, bodying, stabilize overrun	Lambda-	0.10-0.20
Acidified milks:			
Yogurt	Bodying, fruit suspension	Kappa- + locust bean gum	0.20-0.50

(Reproduced with permission from Food colloids, edited by H.D. Graham. Westport, Connecticut, AVI Publishing Co., Inc. (1977), and Handbook of water-soluble gums, edited by R.L. Davidson. New York McGraw-Hill Book Co. (1980))

Table 3

Typical water applications of carrageenan

Use	Function	Carrageenan type	Approx. use level, %
Dessert gels	Gelation	Kappa- + iota- Kappa- + iota- + locust bean gum	} 0.5-1.0
Low-calorie jellies	Gelation	Kappa- + iota- Kappa- + galactomannans	} 0.5-1.0
Pet-foods (canned)	Fat stabilization, thickening, suspending gelation	Kappa- + locust bean gum	0.2-1.0
Fish gels	Gelation	Kappa- + locust bean gum Kappa- + iota-	} 0.5-1.0
Syrups	Suspension, bodying	Kappa- + lambda-	0.3-0.5
Fruit drink powders and frozen concentrates	Bodying Pulping effects	Sodium kappa-, lambda- Potassium calcium kappa-	0.1-0.2 0.1-0.2
Relishes, pizza, barbecue sauces	Bodying	Kappa-	0.2-0.5
Imitation milk	Bodying, fat stabilization	Iota-, lambda-	0.03-0.06
Imitation coffee creams	Emulsion stabilization	Lambda-	0.1-0.2

Table 3 (continued)

Use	Function	Carrageenan type	Approx. use level, %
Whipped toppings (artifical)	Stabilize emulsion, overrun	Kappa-, iota-	0.1-0.3
Puddings (nondairy)	Emulsion stabilization	Kappa-	0.1-0.3
Air-treatment gels	Gelation	Kappa- + iota- Kappa- + galactomannans	2.0-3.5
Toothpastes	Binder	Sodium kappa-, lambda-, iota-	0.8-1.2
Lotions	Bodying, emollient	Sodium kappa-, lambda-, iota-	0.2-1.0
Suspensions (industrial)	Suspension	Iota-	0.2-1.0
Dispersions	Suspension, dispersion	Hydrolyzed kappa-, lambda-, iota-	0.2-0.5
Water-based paints	Suspension, flow control emulsion stabilization	kappa-, + galactomannans, iota-	0.15-0.5

(Reproduced with permission from Food colloids, edited by H.D. Graham. Westport, Connecticut, AVI Publishing Co., Inc. (1977), and Handbook of water-soluble gums, edited by R.L. Davidson, New York, McGraw-Hill Book Co. (1980))

strong position in this application, due, among other factors, to its immunity to degradation by enzymes which attack cellulose gums.

Carrageenan is used in combination with locust bean and guar galactomannans as a gelling agent in pet foods. However extracted carrageenans have now been almost entirely replaced by semi-refined carrageenan from Eucheuma cottonii (Table 4).

Another major application of the gelling properties of carrageenans is in air freshener gels. The demand for carrageenan in this market peaked in the 1970s and has undergone some erosion since then, but still persists at a fairly stable level.

PRODUCTS

Basically there are three types of carrageenan of commercial importance. These are kappa-, iota-, and lambda-carrageenan. Furcellaran may be considered to be an extreme kappa type. These commercial extractives approximate to the limit polysaccharides, their criteria being functionality rather than strict chemical characterization.

Batches of the basic carrageenans are tested for their functional properties and then blended to produce standardized products. Diluents, usually sucrose or glucose, may be added for standardization. Food grade salts, such as potassium chloride or citrate, and other gums, particularly locust bean gum, may be incorporated in the blend to achieve desired functional properties. In all, more than two hundred different carrageenan and furcellaran blends, tailored to meet specific applications, are presently offered by the several manufacturers, as well as blending houses, under their various trade names.

MARKETING

Sales of carrageenans in millions of kilograms and dollars are shown in Table 4. Figures are shown for both extractive and sime-refined carrageenan. The latter is used almost exclusively for the huge pet food market, and its explosive growth in the 1980-82 period reflects the replacement of extractive by semi-refined in pet food formulations. This changeover has now been accomplished and steady, but unexciting, growth is now projected for both types. The growth rate reflects the maturity of the food processing industry which is the staple outlet for carrageenans. Although the number of new food products introduced annually doubled from 1 026 in 1970 to 2 200 in 1985 volume demand does not increase accordingly. Table 5 shows the distribution of extractive sales by end use and by geographical region. Figure 4 shows that a moderate excess of production capacity over sales exists.

Table 4

World market for refined and
semi-refined carrageenan

Carrageenan	Market size 1982		Volume growth, %	Compound volume growth, %/year [a]
	kg MM	$MM	80-82	83-88
Refined	10.8	88.3	-6.1	1.8
Semi-refined	2.4	10.1	600	4.0
Both	13.2	98.4	3.6	2.2

[a] Projected

Table 5

Distribution of carrageenan sales

By end use	Percent	By region	Percent
Dairy	52	Europe	45
Water gel	16	North America	23
Other food	10	Latin America	12
Non-food	22	Far East	20

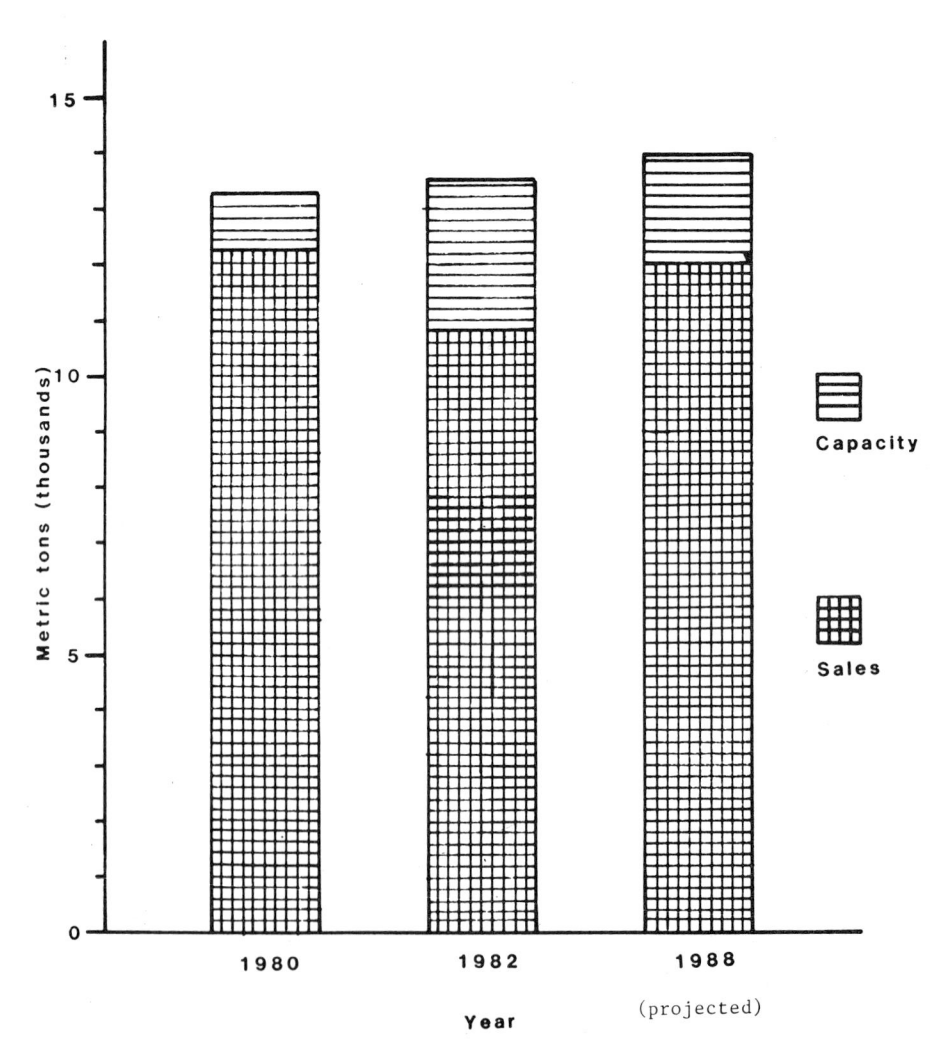

Figure 4 Carrageenan extract capacity vs. sales

Speciality gums, such as carrageenans, are sold on the basis of their functionality in specific applications and not as commodities. For this reason the major carrageenan manufacturers devote substantial portions of their budgets to maintaining active applications and technical marketing groups to serve the ever-changing needs of their customers. There is no room for complacency in the carrageenan industry.

FUTURE PROSPECTS

After nearly fifty years of development the carrageenan industry can now be said to have come of age and to be a mature industry. As previously mentioned its close ties to the food processing industry, likewise in its maturity, presages that future growth should be steady, if unspectacular. New products and applications can be expected, but these may be slow in coming. An insight into the time scale may be gained from the observation that the last "new" application for carrageenan of any great commercial significance was air freshener gels in the early 1970s. Opportunities exist wherever the functionality of carrageenans can confer advantages not possessed by cheaper competitive gums.

A favorable factor has been the stabilization of seaweed supplies and prices due to the advent of Eucheuma farming. Barring political upheavals in the harvesting regions the industry can remain assured of adequate supplies of good quality weed at reasonable prices.

Future progress now appears to lie in the areas of achieving cost reduction in processing and developing more versatile and better quality-controlled products.

REFERENCES

Allsobrook, A.J.R., J.R. Nunn and H. Parolis, 1971. Sulphated polysaccharides of the Grateloupiaceae family. Part 5. A polysaccharide from Aeodes ulvoidea. Carbohydr.Res., 16:71-8

Anderson, N.S., T.C.S. Dolan and D.A. Rees, 1968. Carrageenans. Part 3. Oxidative hydrolysis of methylated kappa-carrageenan and evidence for a masked repeating structure. J.Chem.Soc.(C), 1968:596-601

_____, 1973. Carrageenans. Part 7. Polysaccharides from Eucheuma spinosum and Eucheuma cottonii. The covalent structure of iota-carrageenan. J.Chem.Soc.Lond.Perkins Trans.I, 1973:2173-6

Anderson, N.S., et al., 1968. Carrageenans. Part 4. Variations in the structure and gel properties of kappa-carrageenan, and the characterization of sulphate esters by infrared spectroscopy. J.Chem.Soc.(C), 1968:602-6

_____, 1968a. Carrageenans. Part 5. The masked repeating structures of lambda- and mu-carrageenans. Carbohyd.Res., 7:468-73

Bjerre-Petersen, E., J. Christensen and P. Hemmingsen, 1973. Furcellaran. In Industrial gums, edited by R.L. Whistler and J.N. BeMiller. New York, Academic Press, pp.123-36

Bourgade, G., 1871. Improvement in treating marine plants to obtain gelatine and c. U.S. Patent 112,535.

Chen, L. C-M., et al., 1973. The ratio of kappa- to lambda-carrageenan in nuclear phases of the rhodophycean algae, Chondrus crispus and Gigartina stellata. J.Mar.Biol.Assoc.U.K., 53:11-6

Collins, T.F.X., T.N. Black and J.H. Prew, 1977. Long term effects of calcium carrageenan in rats. 1. Effects on reproduction. Food Cosmet.Toxicol., 15:533-8

_____, 1977a. Long term effects of calcium carrageenan in rats. 2. Effects on fetal development. Food Cosmet.Toxicol., 15:539-45

_____, 1979. Effects of calcium and sodium carrageenans and iota-carrageenan on hamster foetal development. Food Cosmet.Toxicol., 17:443-9

Deslandes, E., et al., 1985. Evidence for lambda-carrageenans in Solieria chordalis (Solieriaceae) and Callibepharis jubata, C. ciliata and Cystoclonium purpureum (Rhodophyllidaceae). Bot.Mar., 28:317-8

DiNinno, V., E.L. McCandless and R.A. Bell, 1979. Pyruvic acid derivative of a carrageenan from a marine red alga (Petrocelis species). Carbohydr.Res., 71:C1-4

Dolan, T.C.S. and D.A. Rees, 1965. The carrageenans. Part 2. The positions of the glycosidic linkages and sulphate esters in lambda-carrageenan. J.Chem.Soc. 1965:3534-9

Doty, M.S., 1973. Farming the red seaweed Eucheuma for carrageenans. Micronesica, 9:59-73

Doty, M.S. and G.A. Santos, 1978. Carrageenans from tetrasporic and cystocarpic Eucheuma species. Aquat.Bot., 4:143-9

Eisses, J., 1952. The research of gelatinous substances in Indonesian seaweeds at the Laboratory for Chemical Research, Bogor. J.Sci.Res.Indon., 1:44-9

―――――――, 1953. Seaweeds in the Indonesian trade. Indon.J.Nat. Sci., 1(2-3):41-56

Greer, C.W. and W. Yaphe, 1984. Characterization of hybrid (beta-kappa-gamma) carrageenan from Eucheuma gelatinae J. Agardh (Rhodophyta, Solieriaceae) using carrageenases, infrared and ^{13}C-nuclear magnetic resonance spectroscopy. Bot.Mar., 27:473-8

Guiseley, K.B., N.F. Stanley and P.A. Whitehouse, 1980. Carrageenan. In Handbook of water-soluble gums, edited by R.L. Davidson. New York, McGraw-Hill Book Co., pp.5-1 to 5-30

Hirase, S. and K. Watanabe, 1972. The presence of pyruvate residues in lambda-carrageenan and a similar polysaccharide. Bull. Inst.Chem.Res.Kyoto Univ., (50):332-6

Hirase, S., C. Araki and K. Watanabe, 1967. Component sugars of the polysaccharide of the red seaweed Grateloupia elliptica. Bull.Chem.Soc.Japan, 40:1445-8

Lawson, C.J. and D.A. Rees, 1968. Carrageenans. Part 6. Reinvestigation of acetolysis products of lambda-carrageenan. Revision of the structure of α-1,3-galactotriose, and a further example of the reverse specificities of glycoside hydrolysis and acetolysis. J.Chem.Soc.(C), 1968:1301-4

―――――――, 1970. An enzyme for the metabolic control of polysaccharides conformation and function. Nature, Lond., 227:390-3

Lawson, C.J., et al., 1973. Carrageenans. Part 8. Repeating structures of galactan sulphates from Furcellaria fastigiata, Gigartina canaliculata, Gigartina atropurpurea, Ahnfeltia durvillaei, Gymnogongrus furcellatus, Eucheuma cottonii, Eucheuma spinosum, Eucheuma isiforme, Eucheuma uncinatum, Agardhiella tenera, Pachymenia hymantophora and Gloiopeltis cervicornis. J.Chem.Soc.Lond.Perkins Trans.1, 1973:2177-82

Lund, S. and E. Bjerre-Petersen, 1952. Industrial utilization of Danish seaweeds. Proc.Int.Seaweed Symp., 1:85-7

Lund, S. and E. Bjerre-Petersen, 1961. Collection and utilization of Danish Furcellaria, 1946-60. Proc.Int.Seaweed Symp., 4:410-1

MacMullan, E.A. and F. Eirich, 1963. The precipitation reaction of carrageenan with gelatin. J.Colloid Sci., 18:526-37

McCandless, E.L. and M.R. Gretz, 1984. Biochemical and immunochemical analysis of carrageenans of the Gigartinaceae and Phyllophoraceae. Hydrobiologia, 116/117:175-8

McCandless, E.L., J.A. West and M.D. Guiry, 1982. Carrageenan patterns in the Phyllophoraceae. Biochem.System.Biol., 10:275-84

―――――――, 1983. Carrageenan patterns in the Gigartinaceae. Biochem.System.Ecol., 11:175-82

Moirano, A., 1977. Sulfated seaweed polysaccharides. In Food Colloids, edited by H.D. Graham. Westport, Connecticut, AVI Publishing Co., pp.347-81

Morris, E.R., D.A. Rees and G. Robinson, 1980. Cation-specific aggregation of carrageenan helices: domain model of polymer gel structure. J.Mol.Biol., 138:349-62

Moss, J.A., 1978. Essential considerations for establishing seaweed extraction factories. In The marine plant biomass of the Pacific Northwest coast, edited by R. Krauss. Corvallis, Oregon State University Press, pp. 301-14

Mueller, G.P. and D.A. Rees, 1968. Current structural views of red seaweed polysaccharides. In Drugs from the sea, edited by H.D. Freundenthal. Washington, D.C., Marine Technology Society, pp.241-55

Nunn, J.R. and H. Parolis, 1968. A polysaccharide from Aeodes orbitosa. Carbohydr.Res., 6:1-11

Painter, T.J., 1966. The location of the sulphate half-ester groups in furcellaran and kappa-carrageenan. Proc.Int.Seaweed Symp., 5:305-13

Paoletti, S., O. Smidsrød and H. Grasdalen, 1984. Thermodynamic stability of the ordered conformation of carrageenan polyelectrolytes. Biopolymers, 23:1771-94

Parolis, H., 1981. The polysaccharides of Phyllymenia hieroglyphica (=P. belangeri) and Pachymenia hymantophora. Carbohydr. Res., 93:261-7

Penman, A. and D.A. Rees, 1973. Carrageenans. Part 9. Methylation analysis of galactan sulphates from <u>Furcellaria fastigiata</u>, <u>Gigartina canaliculata</u>, <u>Gigartina chamissoi</u>, <u>Gigartina atropurpurea</u>, <u>Ahnfeltia durvillaei</u>, <u>Gymnogongrus furcellatus</u>, <u>Eucheuma isiforme</u>, <u>Eucheuma uncinatum</u>, <u>Aghardhiella tenera</u>, <u>Pachymenia hymantophora</u> and <u>Gloiopeltis cervicornis</u>. Structure of xi-carrageenan. <u>J. Chem.Soc.Lond.Perkins Trans.1</u>, 1973:2182-7

_____, 1973a. Carrageenans. Part 10. Synthesis of 3,6-di-<u>O</u>-methyl-D-galactose, a new sugar from the methylation analysis of polysaccharides related to xi-carrageenan. <u>J.Chem.Soc.Lond.Perkins Trans.1</u>, 1973: 2188-91

_____, 1973b. Carrageenans. Part 11. Mild oxidative hydrolysis of kappa- and lambda-carrageenans and the characterization of oligosaccharide sulphates. <u>J.Chem. Soc.Lond.Perkins Trans.1</u>, 2191-6

Pernas, A.J., et al., 1967. Chemical heterogeneity of carrageenans as shown by fractional precipitation with potassium chloride. <u>Acta Chem.Scand.</u>, 21:98-110

Rees, D.A., 1963. The carrageenan system of polysaccharides. Part 1. The relation between the kappa- and lambda-components. <u>J.Chem.Soc.</u>, 1963:1821-32

_____, 1972. Mechanism of gelation in polysaccharide systems. <u>In</u> Gelation and gelling agents. London, British Food Manufacturing Research Association, Symposium proceedings, 13:7-12

Smidsrød, O., et al., 1967. The effect of alkali treatment on the chemical heterogeneity and physical properties of some carrageenans. <u>Acta Chem.Scand.</u>, 21:2585-98

Smith, D.B. and W.H. Cook, 1953. Fractionation of carrageenin. <u>Arch.Biochem.Biophys.</u>, 45:232-3

Smith, D.B., W.H. Cook and J.L. Neal, 1954. Physical studies on carrageenin and carrageenin fractions. <u>Arch.Biochem. Biophys.</u>, 53:192-204

Smith, H.M., 1905. The utilization of seaweeds in the United States. <u>Bull.U.S.Bur.Fish.</u>, 24:169-71

Snoeren, T.H.M., et al., 1975. Electrostatic interaction between kappa-carrageenan and kappa-casein. <u>Milchwissenschaft</u>, 30:393-6

Stancioff, D.J. and D.W. Renn, 1975. Physiological effects of carrageenan. In Physiological effects of food carbohydrates, edited by A. Jeanes and J. Hodge. Washington, D.C., American Chemical Society, pp.282-95

Stancioff, D.J. and N.F. Stanley, 1969. Infrared and chemical studies on algal polysaccharides. Proc.Int.Seaweed Symp., 6:595-609

U.S. Food and Drug Administration, 1979. Carrageenan, salts of carrageenan, and Chondrus extract (carrageenin); withdrawal of proposal and termination of rulemaking procedure. Fed.Reg., 44:40343-5

U.S. Food and Nutrition Board, 1981. Committee on Codex Specification, Carrageenan. In Food chemicals codex. Washington, D.C., National Academy Press, pp. 74-5 3rd.ed.

Zaneveld, J.S., 1955. Economic marine algae of tropical south and east Asia and their utilization. Spec.Publ.IPFC, (3):55 p.

_____, 1959. The utilization of marine algae in tropical south and east Asia. Econ.Bot., 13(2):89-131

CHAPTER 4

PREPARATION AND MARKETING OF SEAWEEDS AS FOODS

by

Kazutosi Nisizawa
Department of Fisheries
College of Agriculture and Veterinary Medicine
Nihon University, 34 1, Shimouma-3
Setagaya-ku, Tokyo 154
Japan

BACKGROUND

Seaweeds have been used as a human food since ancient times, particularly in the region bounded by China, the Korean Peninsula and Japan. In Japan, remains of marine algae, which are eaten at present, have often been found in the relics of aborigines, mixed with shells and fish bones, even in the Jomon-pattern era (B.C. 300-6 000). It has been formally recorded that the Yamato Imperial court which had been founded in the beginning of the fourth century, made the inhabitants pay taxes using various edible seaweeds such as nori (Porphyra), wakame (Undaria), hiziki (Hizikia) and konbu (Laminaria). These seaweeds were also offered to the spirits of ancestors in the rituals of the court (Miyashita, 1974). In China, the use of Porphyra and its value as a food were recorded in a book published between 533-544 AD. In the Sung Dynasty (960-1279 AD) Porphyra, from Haitan Island of Fujian Province, was presented every year as a special delicacy to the Emperor (Tseng, 1984).

Varieties of brown, red and green seaweeds find use as foods although the browns and reds predominate. Species of Laminaria, Undaria pinnatifida and Hizikia fusiforme are the major brown seaweeds used; they are usually sold as dried products but a wet, salted form of Undaria is very popular. Konbu is made from several species of Laminaria, often referred to by the general name of kelp; it is boiled as a vegetable, used as a soup stock, as a seasoning for rice dishes and eaten as a snack food.

Wakame, from Undaria pinnatifida is, when reconstituted, one of the softest brown seaweeds and is widely used in soups. Hiziki, made by boiling Hizikia fusiforme and sun-drying, is usually cooked with soybean or fried bean curd. Many species of the red seaweed, Porphyra, are used to make the thin dried sheets of Laver which are used widely in the oriental cuisine. The product form Porphyra is also named purple laver; this distinguishes it from green laver which is produced in smaller quantity from several green seaweeds such as species of Enteromorpha, Monostroma and Ulva. Laver has a much higher protein content (25-35%) than the products from brown seaweeds (8-15%) and, nutritionally, is the superior of these foods.

The major proportion of the seaweed used in China, Korea and Japan is purchased as a dried product. However there is also a market in some varieties of fresh seaweeds which are used as a salad vegetable or as garnishes for other dishes such as fish. Species of Caulerpa, Eucheuma and Gracilaria are used for this purpose, especially in some of the warmer southeast Asian countries such as the Philippines, Malaysia, Thailand and Indonesia. Usually naturally-growing species are collected and sold fresh in local markets. In Cebu, Philippines, Caulerpa racemosa is cultivated in brackish-water ponds and the fresh seaweed is sent by air to markets in Manila.

On a worldwide basis, Japan has one of the highest per caput consumptions of seaweed as food so the description here is focused on the processes and products used in this country.

MAIN SEAWEEDS USED AS FOODS

PURPLE LAVER, NORI

1. SOURCES

The purple laver, Porphyra genus, consists of a large taxonomic group which covers nearly 50 species in the world and about 20 species are found in Japan. The individual species are more or less morphologically different, but a fairly large variation is found even in a single species. The shape, the size, the thickness and the colour of their thalli are changeable, depending on the environment in which they grow. In addition, they are considerably different in the individual patterns of variation, so it is very difficult to identify them merely on the basis of external appearance.

The purple laver was first cultivated in Japan as early as the 17th century, the beginning of the Edo era. P. tenera was cultivated until about 30 years ago, but now more than 80% of this has been substituted by P. yezoensis. This is mainly because the latter laver is more strongly resistant to various laver diseases. In China, the cultivation has been developed during the past several years using a strain of Porphyra tenera. In Korea, almost all the nori is obtained from marine culture using both Porphyra tenera and Porphyra yezoensis. A small quantity of natural material (called iwanori in Japanese) is collected.

Porphyra tenera and Porphyra yezoensis are both monoecious but some species are dioecious. P. tenera grows widely in Japan (on the

Pacific side of Honshu, in the Inland Sea, on the eastern and northern coasts of Kyushu) and on every coast of Korea and China. P. yezoensis is also distributed widely on the coasts from western to southern Hokkaido and in the northern areas of the Pacific side of Honshu and in the Korean Peninsula.

2. PREPARATION OF FOOD PRODUCTS

The purple laver fronds grown in the winter season are harvested mechanically from culture grounds and transferred by boats to factories belonging to individuals or to groups of fishermen. The algal fronds are first washed thoroughly with fresh seawater in a machine to remove any contaminants and epiphytes. They are then cut into pieces with a chopper and a given amount of the thalli suspension is poured into a wooden rectangular frame with a perforated base which allows the water to escape but retains the seaweed. In this way rectangular sheets of laver are formed. The wet sheets are then dried on the frames in a hot-air chamber ($40°C \pm 1°C$) to give sheets of the final product, hoshi nori (dried laver thalli). A sheet of the product is 21 cm long, 19 cm wide and an average weight is about 3.0 g (see Figures 1 to 6).

In Japan the legal cultivation of nori is restricted to those who have been granted licences. There were approximately 45 000 fishermen engaged in this work in 1964 but this has decreased markedly, to around 20 000 people at present. Part of the hoshi nori is processed to toasted and/or seasoned laver products. Recently, some modern foods such as instant soup, jam and wine, which are made from materials mixed with nori, have come onto the market. The annual production of hoshi nori in Japan during recent years is shown in Table 1, the figure for 1984 being 87×10^8 sheets. The quantities exported from Japan are shown in Table 2. The export of hoshi nori was no more than 0.75% of the total production in 1984, but it has increased during recent years with the largest amount going to Taiwan, followed by the USA (Natural Federation of Nori and Shell Fishes Cooperative Associations, 1985). For China, the yield in 1985 amounted to more than 41×10^8 sheets when converted to the Japanese scale of measurement. Korean production is at about the rate of $30-40 \times 10^8$ sheets per year.

3. NUTRITIVE VALUE

Hoshi nori has a vitamin content almost equal to fresh laver because it has been manufactured under controlled conditions. It is high in protein content and the lipid content is also rather high relative to that of other seaweeds. These figures are shown in Table 3 (Science and Technology Agency, 1982; Noda, 1971, 1981; Horiguchi, Noda and Naka, 1971; Kanazawa, 1963). It is clear that hoshi nori is generally rich in vitamins, and it is particularly worthy of note that it contains almost 30 micrograms percent of

Figure 1　Secondary growing nori thalli (Porphyra yezoensis) in the maricultural ground

Figure 2　Harvest boat of nori with a harvesting machine of the cut and suction type

Figure 3 The hoshi nori is baked in a toasting machine to make a toasted-nori product which has a characteristic baked flavour

Figure 4 The toasted nori is cut, in some cases, into a small square sheet (about 8 x 8 cm). Several sheets are wrapped in a cellophane bag

Figure 5 Drying nori sheets (21 x 19 cm) in a hot air drier (40°C + 1°C) with conveyor system. The product is hoshi nori

Figure 6 Each nori-sheet bundle (tied with a white tape), consists of 10 piles (with 10 sheets of hoshi nori per pile). They are arranged on the table after examination by an authorized inspector (right side of front) before distribution to customers

Table 1

Annual production of hoshinori from 1980 to 1984 in Japan

	1980	1981	1982	1983	1984
Harvest (x 10000 sheets)	777800	766700	722700	1041600	867300
Value (x 10000 yen)	11060200	9744300	12720500	12714100	11171800

Table 2

Export of various processed nori products from Japan during the past several years (kg)

Country	1981 Dried	1981 Toasted, Seasoned	1982 Dried	1982 Toasted, Seasoned	1983 Dried	1983 Toasted, Seasoned	1984 Dried	1984 Toasted, Seasoned
Taiwan	6100	263	246188	10611	346152	20028	663240	13502
Hong-Kong	3300	44105	29468	107104	40	84736	420	69610
Saudi Arabia	7728	370	21856	682	1820	264	600	558
England	624	256	1352	349	836	619	1710	399
Belgium	512	26	484	26	520	103	492	70
France	548	709	192	219	364	303	632	660
Italy	448	89	116	74	308	35	16	98
Canada	3948	870	4848	1578	1508	2453	2364	3080
USA	98300	41732	115844	56327	81264	72223	103844	101661
Brazil	180	1290	652	160	2408	78	4160	561
Australia	536	479	1164	164	784	963	544	2902
Korea	273532	49930	614912	22349	8364	3834	6440	1505

Vitamin B_{12}, which is roughly comparable to that in animal viscera. The contents of free and bound amino acids are shown in Table 4 (Kagawa, 1983).

Both the free and bound amino acids of nori are rich in alanine, aspartic acid, glutamic acid and glycine which represent the characteristic taste of nori (Noda, et al., 1981), being in synergistic action with the taste of inosinic acid which also occurs in this alga (Nakamura, et al., 1968). It is most striking that nori contains taurine in a remarkably high amount, more than 1 200 mg% of dry weight (Noda, Horiguchi and Araki, 1975). This sulfated amino acid is known to be effective for liver activity in preventing the occurrence of gallstone disease and in controlling blood cholesterol level (Tsuji, et al., 1981, 1983). In addition, because the content of arginine is fairly high in nori, there is a possibility that taurine and arginine might combine after intake into the human body to form taurocyamine which has been reported to show a potent activity in preventing high cholesterol levels (Hiramatsu, Niitani and Mori, 1981).

A sheet of hoshi nori also contains 30-45 mg% of eicosapentaenoic acid (EPA) which has been reported to prevent atherosclerosis (Dyerberg, et al., 1978; Dyerberg and Bang, 1979). Nori has been reported to contain a substance, porphyosin, which showed activity in preventing gastric shay ulcer of rats (Sakagami, 1983). When various samples of marine algal powder or their extracts were given to rats, mixed with their basic diet, it was found that powder of nori was most effective in preventing 1,2-dimethylhydrazine-induced intestinal carcinogenesis of rats (Yamamoto and Maruyama, 1985).

4. MARKETING

The raw fresh fronds of nori, and hoshi nori, are subjected to marketing routes as shown in Figure 7.

5. PRICE

In Japan the price of hoshi nori at the Local Fisheries Cooperative Association (LCFA) in 1985 has been reported to be 3 823 yen per kg (US$ 23.9 assuming US$ 1 is equivalent to 160 yen). The price, however, rose to 8 333-10 000 yen by the time the product reached consumers via the normal marketing channels. According to informed circles, the price of hoshi nori in China was about 670 yen per kg in 1985. In Korea the price is estimated to be about the level of 2 300 yen per kg.

6. COOKING

Hoshi nori is usually sold as a set of ten sheets packed in a bag. This amount is approximately 30 g and is called a "joo". It is

Table 3

Nutritive composition data from several different samples of hoshinori on a dry weight basis

Main composition (g%)		Mineral (mg%)		Vitamin (mg%)	
Protein	43.6	Ca	440	A (Potency)	15000 IU
Lipid	2.1	P	650	B_1	1.29
Carbohydrate Nonfibrous	44.4	Fe	13.0	B_2	3.82
		Na	570	Niacin	11.0
Fiber	2.0	K	2400	B_6	1.04 [c]
Ash	7.8	Mn	2 [a]	B_{12}	0.029 [c]
		Zn	10 [a]	Cholin	292 [c]
		Cu	1.47 [a]	Inositol	6.2 [c]
		Se	0.08 [b]	C	112.5

[a] Noda, 1971; Noda et al., 1981
[b] Horiguchi, Noda and Naka, 1971
[c] Kanazawa, 1963; and for the rest see Standard Table, 1982

Table 4

Free and bound amino acids in hoshinori (g%) on a dry weight basis

Amino acid	Free	Bound (total N x 6.25 as protein)	Amino acid	Free	Bound (Total N x 6.25 as protein)
Alanine	1528	9.92	Methionine	2	3.36
Arginine	15	5.92	Phenylalanine	7	5.28
Aspartic acid	322	8.48	Proline	4	4.64
Glutamic acid	1330	9.28	Serine	37	4.80
Glycine	24	6.81	Threonine	46	3.20
Histidine	10	1.18	Tryptophane	Trace	1.10
Isoleucine	20	4.00	Tyrosine	13	2.40
Leucine	31	7.68	Valine	15	9.28
Lysine	12	2.56	Taurine	1210	-

a/ Noda, Horiguchi and Araki, 1975

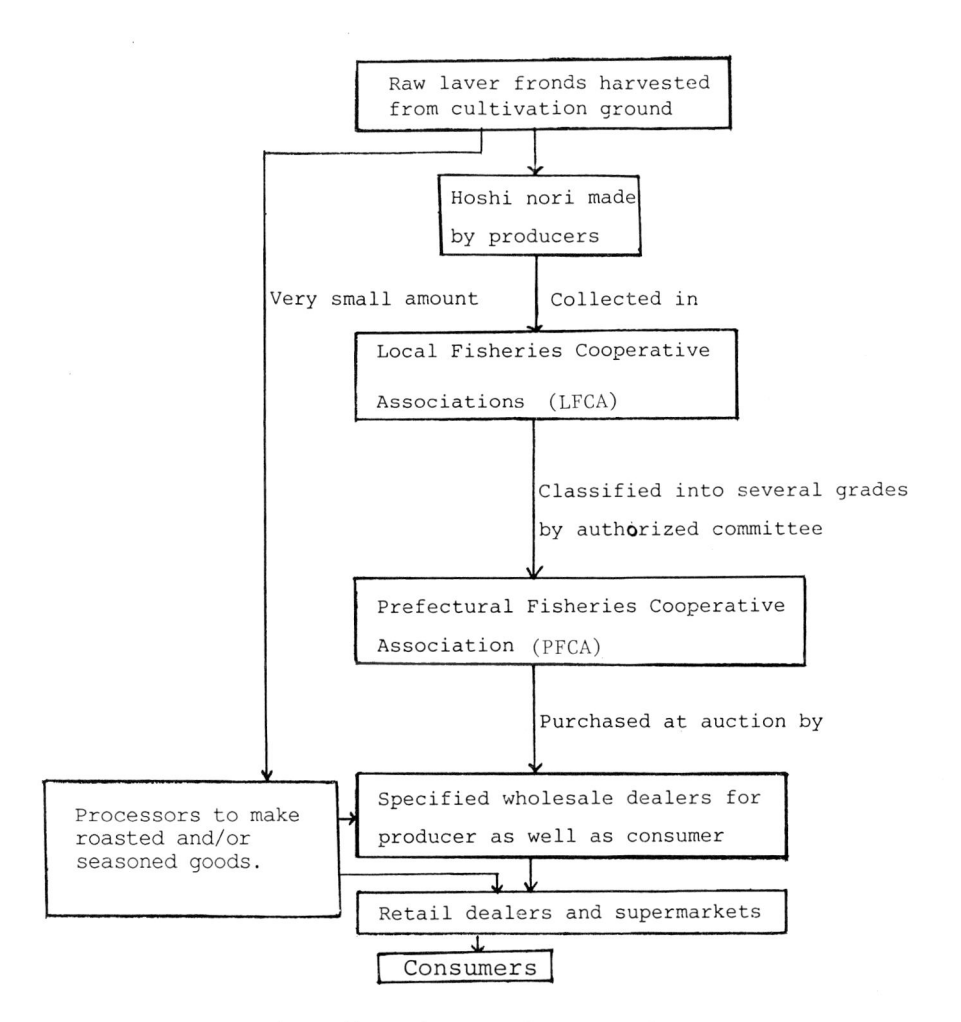

```
         ┌─────────────────────────────┐
         │ Raw laver fronds harvested  │
         │ from cultivation ground     │
         └─────────────────────────────┘
                       │
                       ▼
              ┌──────────────────┐
              │ Hoshi nori made  │
              │ by producers     │
              └──────────────────┘
   Very small amount │ Collected in
                     ▼
              ┌──────────────────────────────┐
              │ Local Fisheries Cooperative  │
              │ Associations (LFCA)          │
              └──────────────────────────────┘
                       │ Classified into several grades
                       │ by authorized committee
                       ▼
              ┌──────────────────────────────┐
              │ Prefectural Fisheries Cooperative │
              │ Association (PFCA)           │
              └──────────────────────────────┘
                       │ Purchased at auction by
                       ▼
┌─────────────────┐  ┌────────────────────────────────┐
│ Processors to make │→│ Specified wholesale dealers for │
│ roasted and/or    │ │ producer as well as consumer    │
│ seasoned goods.   │ └────────────────────────────────┘
└─────────────────┘           │
                              ▼
              ┌──────────────────────────────────┐
              │ Retail dealers and supermarkets  │
              └──────────────────────────────────┘
                              ▼
                     ┌─────────────┐
                     │  Consumers  │
                     └─────────────┘
```

Figure 7 Marketing of nori products

used mainly as a luxury food after being slightly baked. Sushi is a typical Japanese food which consists of a small mass of soured, boiled rice with a topping on it, often a slice of various kinds of raw fish. Nori is sometimes wrapped around the outside of the rice ball which then usually has a tuna (tunny) slice inserted in it (Figure 8).

In other uses, nori is cut into small pieces, after a short baking, and sprinkled over boiled rice and noodles. It is also processed to tsukudani nori, a preserved food boiled down in soy sauce; and used as an appetizing luxury food, although most tsukudani nori is made from an aonori, Monostroma. Recently nori has been used as the raw material for jam and wine, the products being called nori jam and nori wine, respectively.

In China, most nori products are used for soup and for seasoned foods by frying in a hot pan. In Korea, most of the products are used for soup but some are cooked according to Japanese or Chinese styles; a popular method is to fry hoshi nori in a hot pan with a little oil.

GREEN LAVER, AONORI

1. SOURCES

Aonori is the commercial name of a mixture of several green seaweeds such as sea lettuce (Ulva), genuine green laver (Enteromorpha) and Monostroma. Of these green seaweeds, Monostroma latissimum and Enteromorpha prolifera are cultivated for commercial purposes at present, although the former comprises nearly 90% of the cultivated products.

Monostroma latissima is distributed widely in most bays and gulfs on the Pacific side of the cental districts of Japan. This green laver takes its habitat generally at upper parts of the intertidal zone. Enteromorpha prolifera occurs widely in most parts of the Japanese, European and North American coasts, and grows mainly in estuaries just below low-tide water level.

2. PREPARATION OF FOOD PRODUCTS

The green laver fronds are dried in the sun or in a dryer immediately after harvesting. Some of the aonori is brought to market in this form and is used as a sprinkling powder or condiment on cooked rice. Only the fronds of Monostroma are processed as a preserved food, by boiling down with soy sauce and sugar. The recent yields of aonori in Japan are shown in Table 5 (Anon., 1984). Of the total aonori produced, more than 90% comes from Monostroma. Part of the dried aonori is sold in shops or supermarkets, packed in a bag of plastic film of an appropriate size. In Korea recently, the production of aonori (mainly Monostroma) has begun to increase and now amounts to 5 000 metric tons, wet weight, per year. No estimates are available from China.

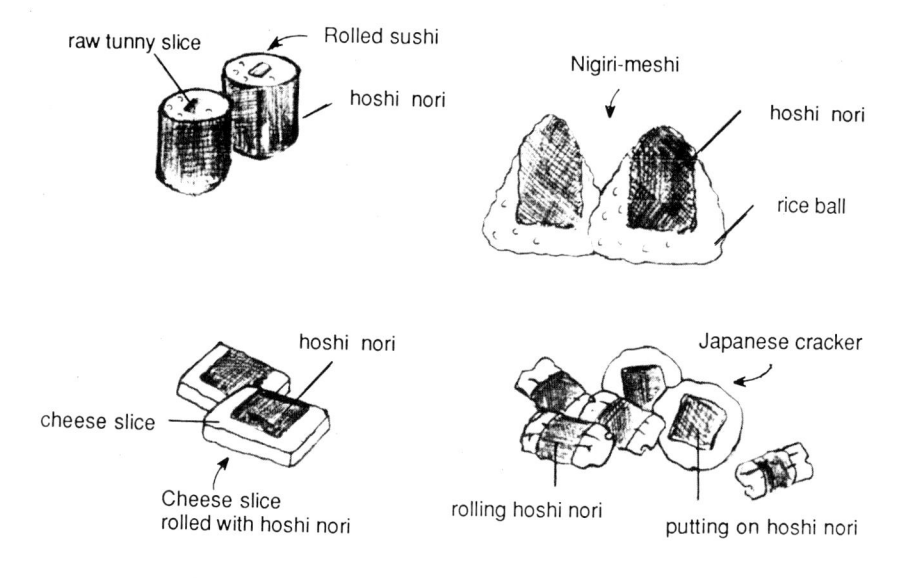

Figure 8 Using nori

Table 5

Annual production of aonori (Monostroma and Enteromorpha) in Japan

Year	1981	1982	1983	1984
Harvest kg (dry wt.)	824135	995929	11094345	1098509

3. NUTRITIVE VALUE

The main composition of the three kinds of green seaweeds is shown in Table 6 and the vitamin and mineral content are given in Table 7 (Science and Technology Agency, 1982). Note that these seaweeds contain 20-26% protein and 15-23% mineral. Compared with the others, Enteromorpha has a low sodium content but is high in calcium and iron. The amount of the vitamin B group is generally higher than that of common vegetables, particularly in the case of Enteromorpha. In contrast, while Enteromorpha has the highest vitamin A content among the three seaweeds, it is approximately only half as much as that of spinach.

These green seaweeds have been found to contain some compounds with pharmacological activity. Water-soluble sulfated polysaccharides, which show activity for the anticoagulation of blood, have been found in all three (Sekiguchi and Maeda, 1983). Monostroma has been reported to contain several kinds of betaines, of which β-homobetaine shows a high activity for lowering the artificially elevated level of plasma cholesterol in rats (Abe and Kaneda, 1975). The dialyzed residue of hot water-extractable substances from Monostroma have been demonstrated to increase the life span of male mice implanted with L.1210 leukemia cells (Yamamoto, et al., 1982). Further, Monostroma contains nearly 200 mg% (on a dry weight basis) of a compound which has been identified as dimethyl β-propiothetin (Noda, 1983, unpublished observation). This compound has been found to prevent gastric ulcer of guinea pigs as well as being useful as a tonic for liver function (Matano, 1959). Dimethyl sulfide, which emits a flavour typical of marine algae, is enzymatically formed from this thetin.

4. MARKETING

The Monostroma fronds, harvested by fishermen, are processed in a similar way to that described previously for the production of hoshi-nori from Porphyra. The resulting square-shaped sheets are dried in the sun or a dryer. Thus, the algal products are collected in the Local Fisheries Cooperative Association and classified, depending on the quality, after examination by authorized inspectors. Then, they are sold by auction at the Prefectural Fisheries Cooperative Association to the wholesale dealers, from whom various retail dealers and processors buy the products. In Japan, Monostroma latissimum is mainly cultivated in Mie Prefecture.

5. PRICE

In Japan hoshi aonori, which includes a large part of Monostroma and a minor part of Enteromorpha, is delivered to wholesale dealers at an average cost of 1 600 yen per kg. In Korea, aonori costs about 1 200 yen per kg, roughly half the cost of Korean hoshi nori.

Table 6

Main composition of dried green algae

Alga	Protein	Lipid	Carbohydrate Nonfibrous	Fiber	Ash
Ulva sp (Sea lettuce)	26.1	0.7	46.1	5.1	22.6
Enteromorpha sp (Green laver)	19.5	0.3	58.1	6.8	15.2
Monostroma sp	20.0	1.2	57.2	6.7	14.9

Table 7

Mineral and vitamin of dried green algae

Alga	Ca	P	Fe	Na	K	A (IU)	B_1	B_2	Niacin	C
Ulva sp (Sea lettuce)	1120	94	6.2	3183	731	590	0.08	0.57	11.8	12.0
Enteromorpha sp (Green laver)	910	800	35.0	570	3500	13000	0.60	2.05	6.5	43.2
Monostroma sp (Hitoegusa)	690	200	2.5	1800	810	2700	0.43	1.33	3.5	54.0

Mineral (mg%) — Ca, P, Fe, Na, K, A; Vitamin (mg%) — B_1, B_2, Niacin, C

KONBU (LAMINARIA)

1. SOURCES

Japanese Laminaria and related species used so far for foodstuffs are collectively called konbu and this includes the following species: Laminaria japonica, L. japonica var. ochotensis, L. angustata, L. angustata var. longissima, Arthrothammus bifidus, Kjellmaniella gyrata and K. crassifolia.

All these species grow mainly in particular areas of the coast of Hokkaido. For example, L. japonica is distributed mainly along the coast on both sides of Hakodate, L. angustata is distributed mainly on both sides of Hidaka and L. japonica var. ochotensis is distributed on both sides of Rishiri, this last city being the northernmost place in Hokkaido. Recently, konbu has been successfully cultivated and it has even been planned to utilize it as an energy resource, in the same way as has been planned for Macrocystis in USA.

Annual yields of konbu in Japan during recent years are shown in Table 8 of the Statistical Year Book, 1983 (Statistics and Information Department, 1985). The yield of konbu from marine aquaculture has increased gradually and the total amount in 1983 was more than 170 000 t of wet weight (roughly 35 000 t of dry weight). Of this amount of konbu, L. angustata var. longissima accounts for 30% of the total, L. japonica 20%, L. angustata 15% and L. japonica var. ochotensis 10%. The residual 25% is covered by other species. China is the most advanced country in the cultivation of Laminaria; it produces about 230 000 metric tons per year (1983) (dry weight, containing about 20% moisture). Roughly 35% of the konbu in China is used as raw material for alginate production. It has been reported that about 1 400 t (dry weight) of Chinese konbu were imported to Japan in 1983. In Korea, the production in 1982 was 761 metric tons of natural kelp and 3 987 metric tons of cultivated kelp, both quantities being wet weight; however the latest figures, for 1984, show the collection of natural kelp has fallen to 34 metric tons while cultivated kelp has risen to 7 927 metric tons (Korea, Ministry of Agriculture and Fisheries, 1983 and 1985).

2. PREPARATION OF FOOD PRODUCTS

A. Suboshi konbu

Konbu is commonly dried in the sun on ground covered with pebbles, but is dried in a dryer on a cloudy or rainy day. The product is called suboshi konbu, and is different in quality in each fishing ground, mainly because of the different species collected. The suboshi konbu, depending on its source and consequent properties, is processed into various products.

B. Salted konbu

Fronds of suboshi konbu of a high quality are cut into square pieces (2 cm x 2 cm) or rectangles (2 cm x 4 cm). They are boiled with seasoning or soy sauce, mirin (a kind of sweet Japanese sake) and sugar until most of the water has evaporated, and then finished by drying. The product is called shio konbu, and is mainly made from Laminaria japonica.

C. Salted and boiled down konbu

The suboshi konbu, mainly of Laminaria angustata origin, is boiled down with soy sauce and some seasonings until the broth has been almost evaporated, and a salty plastic product of konbu obtained. This is called tsukudani konbu.

D. Shredded konbu

The suboshi konbu is shredded into strings and soaked in soy sauce to make a konbu pickle such as mitsumame konbu which is often seasoned with soy after mixing with shredded dried cuttlefish and herring roe. The shredded product is called kizami konbu.

E. Sliced konbu

The suboshi konbu, mainly of Laminaria japonica origin, is softened by dipping in vinegar and then sliced with a plane from either side of the blade. The thin, wide slices which result are called oboro konbu or filmy konbu (Figure 9). The residual middle part of the blade which is slightly thicker and light yellowish is called battera konbu. The latter is a more transparent product than the former.

F. Rolled konbu

The suboshi konbu, mainly from Laminaria angustata and Laminaria angustata var. longissima, is cut into rectangles (5 cm x 10 cm) and rolled like a paper roll. After it has been bound with a string of dried gourd shavings, it is boiled down with a condiment consisting of soy sauce and sugar. The product is called konbumaki (Figure 11).

G. Konbu tea

Blades of Laminaria japonica of high quality are dried thoroughly in hot air and powdered mechanically. A small amount of salt and sugar is added, as required, and it is used in the same way as green tea powder which is normally used in the Japanese tea ceremony.

Table 8

Recent annual yields of konbu in Japan
(tons on a wet weight basis)

Year	Marine Fisheries	Marine Aquaculture	Total
1975	157760	15696	173456
1980	124816	38562	163378
1983	129043	44345	173388

Figure 9 The oboro konbu (sliced konbu blade) is made by this type of machine. The products are hung in the same room, looking like long tapes.

II. Others

Konbu jam and konbu sake have been manufactured using the konbu as part of the raw materials. Tablets made of konbu, and konbu noodles in which konbu powder is mixed with wheat or buckwheat powder, are now also to be marketed.

3. NUTRITIVE VALUE

The main constituents of several types of suboshi konbu, and the products processed from them, are shown in Table 9 (Science and Technology Agency, 1982). In contrast to suboshi konbu, the main composition of processed products is quite variable in amount, suggesting that some constituents may be lost during processing.

Konbu is generally rich in nonfibrous carbohydrate, most of which is alginic acid and fucoidan. A water-extractable mucous mass from konbu consists mainly of fucoidan, water-soluble salts of alginic acid and small amounts of protein. It has been demonstrated by various workers that fucoidan and water-soluble alginate exhibit antibloodcoagulation, antilipemia and antitumor activities in animal experiments (Mori, et al., 1982; Fujiwara, et al., 1984; Yamamoto, et al., 1984).

The amino acid composition of konbu is shown in Table 10 (Kagawa, 1983). Note that free aspartic and glutamic acids, particularly the latter, are conspicuously high, a reflection of the use of konbu as a seasoning material for such a long time in Japan. In addition, there is an amino acid, named laminine, in konbu (Takemoto, Daigo and Takagi, 1965); it has been identified as (5-amino-5-carboxypentyl)-trimethylammonium hydroxide. It shows a transitory action in depressing the blood pressure of experimental animals (Ozawa, Gomi and Otsuki, 1967).

The vitamin, mineral and sterol content of konbu is shown in Table 11 (Standard Tables, 1982). From the nutritive point of view, konbu contains relatively large amounts of fucosterol as well as some essential microelements. The fucosterol may help to prevent the formation of thrombii because it has recently been found to activate a plasminogen activator in cultured cells from the bovine carotid artery (Shimonaka, et al., 1984). Some of the microelements are important for nutrition because they are concerned directly or indirectly with enzyme function. For example, selenium is an active group of glutahione peroxidase and copper constitutes the active center of cytochrome oxidase ($a+a_3$) together with the porphyrin nucleus.

When the fatty acids in konbu are considered, it cannot be overlooked from the nutritional viewpoint that eicosapentaenoic acid (EPA) accounts for 20-25% of the total fatty acids, though the absolute amount is not so large.

Table 9

Main composition of several species of Laminaria and their various products (g% on a dry weight basis)

Sample	Water	Protein	Lipid	Carbohydrate Nonfibrous	Fiber	Ash
Suboshi konbu						
Laminaria angustata	11.6	8.7	2.2	63.5	7.0	18.7
L. angustata var. longissima	10.0	9.2	1.7	50.7	14.3	24.1
L. japonica	9.5	9.1	1.3	64.3	3.7	21.7
L. japonica var. ochotensis	10.0	8.9	2.2	60.3	6.0	22.6
Processed konbu						
Shred (kizami) konbu	29.0	7.6	0.7	53.4	6.5	31.8
Sliced (oboro) konbu	28.5	9.1	1.3	64.3	3.6	21.7
Salted (shio) konbu	21.0	21.3	0.5	51.1	2.5	24.6
Salted and boiled down (tsukudani) konbu	60.4	13.4	2.3	40.4	3.3	38.1
Rolled konbu (konbu maki)	61.7	6.3	1.0	79.4	2.9	10.4
Konbu tea (cha)	2.0	5.8	0.4	40.4	0.4	55.0

Table 10

Free and bound amino acids of konbu (L. japonica)
mg % on a dry weight basis

Amino acid	Free		Bound
	First grade [a]	Second grade [a]	(Total N x 6.25 as protein)
Alanine	150	188	6.90
Arginine	-	-	3.47
Aspartic acid	1450	1775	12.15
Glutamic acid	4100	4226	22.61
Cysteine	-	-	2.00
Glycine	9.2	10.9	4.47
Histidine	0.8	1.6	1.24
Isoleucine	7.5	13.0	3.57
Leucine	5.0	9.9	6.11
Lysine	5.0	10.4	2.75
Methionine	3.1	3.5	1.71
Phenylalanine	4.6	8.7	4.17
Proline	175.0	279.1	5.21
Serine	27.4	37.4	3.52
Threonine	16.7	35.3	2.69
Tryptophane	0.4	0.5	1.45
Tyrosine	4.1	4.9	2.78
Valine	3.1	3.8	6.31
Laminine	6.43 [b]		

[a] This is not an experimental evaluation but is based mainly on the sense of the inspectors; thicker and wider blades with glossy and dark colour are classified in the first grade in general

[b] Nonproteinous amino acid (data from L. angustata)

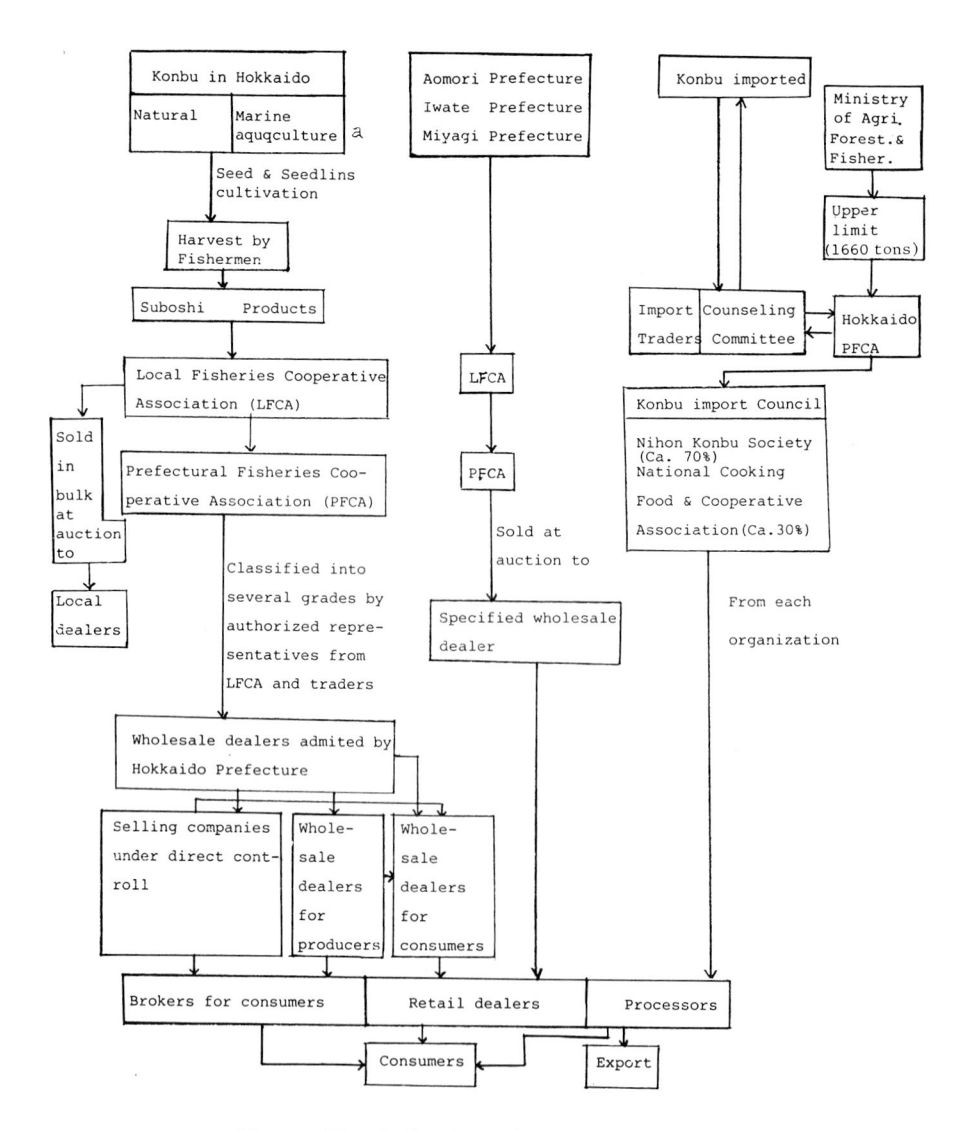

Figure 10 Marketing of konbu products

Table 11

Sterols, vitamins and minerals of konbu (L. japonica)
Data for spinach are listed for comparison

	Konbu	Spinach
Sterol (mg%)		
24-methylene cholesterol	7	-
Fucosterol	9 (170)[f]	-
Vitamin (mg%)		
Carotene	1.1	36.0
Retinol potency (IU)	622	20000
B_1	0.53	0.7
B_2	0.41	1.3
Niacin	1.6	3.0
Folic acid[a]	0.14	-
B_{12}[b]	0.0033	-
C	28	450
Mineral (mg%)		
Ca	789	
P	222	
Na	3111	
K	6778	
Mg[c]	757	
I[d]	250	
Fe	43	
Cu[c]	0.2	
Zn[c]	1.3	
Mn[c]	0.4	
Se[e]	0.4	

[a] Yamada, 1985
[b] Kanazawa, 1963
[c] Innan et al., 1985
[d] Takahashi, 1933
[e] Hosogai, Naoi and Okada, 1978
and others: Standard Tables (Science and Technology Agency, 1982)
[f] Recent unpublished data from one year-growth blade is around 170 mg%

4. MARKETING

It is mainly suboshi konbu that is first prepared from the raw material soon after harvest. The products are cut and folded into roughly one metre lengths by producers and are taken to Prefectural Fisheries Cooperative Associations where they enter the marketing circulation pattern shown in Figure 10.

5. PRICE

In Japan the price of suboshi (sun dried) konbu is very unstable and differs depending on the place where it is harvested. For example, in the case of mitsuishi konbu (<u>Laminaria angusta</u>), the price at the Local Fisheries Cooperative Association is 1 200-1 600 yen per kg and consumers buy it for about 3 000 yen. The price of Chinese suboshi konbu, exported to Japan in 1984, was about 600 yen per kg.

6. COOKING

Suboshi konbu purchased from the market is cooked with konnyaku (devil's tongue), various ground fish pastes, tubers and tuberous roots, etc., to make Japanese hotchpotch, one of the typical Japanese foods which is eaten after simple cooking (see Figure 11). Sometimes the konbu is cooked in the form of a knot, sometimes without any deforming. Suboshi konbu is used for everyday food in such dishes as konbu maki, a seasoned and cooked konbu surrounding dried herring or sliced salmon. These kinds of konbu foods are also sold ready-made. Sliced and seasoned konbu such as battera are used as ingredients in sushi. Nowadays, konbu powder is sometimes mixed in noodles to enhance the nutritive value.

Most konbu used as food in China is put on the market as the suboshi product and it is normally cooked in soup mixed with other foods. Some is cooked in everyday dishes but slightly differently from the Japanese style. In Korea the raw or sun dried (suboshi) konbu is locally used as food, mostly without further treatment.

HIZIKI (HIZIKIA)

1. SOURCES

Hiziki is one of the typical seaweed foods in Japan and the products are manufactured exclusively from <u>Hizikia fusiforme</u>. The alga grows, usually as a large community, just under the tidal zone and is distributed widely on the southern coast of Hokkaido, the whole shore of Honshu, on the Korean peninsula and on most coasts of the China Sea. The annual yields of hiziki in Japan have been reported to be 2 183 t (dry weight) in 1955 and 2 357 t in 1956. It gradually increased until 1959 but began to fall thereafter to 1 211 t in 1960. Domestic harvest in 1985 was around 3 000 t of raw material (1 500 t as commercial product) (Okubo, 1985). In addition, 4 000-5 000 t (2 000-2 500 t as commercial product) were imported from Korea in 1985. In Korea, the major amount of hiziki is the naturally growing material but part of it, corresponding to 100-200 t of commercial product, is cultivated.

2. PREPARATION OF FOOD PRODUCTS

A. Suboshi hiziki

After harvest, the whole alga is washed thoroughly with seawater and dried in the sun to make suboshi hiziki. This product, however, cannot be used as food directly because it has an astringent taste due to the presence of large amounts of phlorotannin. The suboshi hiziki, therefore, is used as a raw material for secondary products.

B. Boiled and sun-dried hiziki (hoshi hiziki)

The suboshi hiziki is boiled in an iron cauldron for 4-5 hours in the presence of the fronds of Eisenia bicyclis (about 1/10 of the amount of hiziki) and then subjected to steaming for 4-5 hours to remove astringent substances. The pigments characteristic of the hiziki products are lost to a large extent during boiling, but they are supplemented by the pigments extracted from Eisenia fronds. The boiled hiziki is then cut into short pieces (about 5 cm) and again dried in the sun. The product is called hoshi hiziki (Figure 12) and it is distinguished from suboshi hiziki. To obtain a marketable product, the hoshi hiziki is normally processed by seasoning with soy sauce and sugar mixed with boiled soybean or fried beancurd.

3. NUTRITIVE VALUE

Figures for the main constituents of hiziki, and other analytical data, are shown in Table 12 (Science and Technology Agency, 1982). Note that while the content of niacin is relatively high, most of the other vitamins are considerably reduced during the boiling down treatment; it still contains a relatively large amount of fucosterol. Hiziki also contains relatively large amounts of microelements, being particularly rich in iron and copper. Although the total lipid content is very low, EPA accounts, as with konbu, for 20-25% of the total fatty acids in hiziki.

4. MARKETING

The sun-dried hiziki (suboshi hiziki) is collected in the Local Fisheries Cooperative Associations and sold by auction directly to wholesale dealers (not via the PCFA, as in Figure 10). The wholesale dealers later sell to the processors who prepare the hoshi hiziki from the suboshi hiziki by the boiling treatment described above. Most of the hiziki is sold in this form. Only a small fraction of it is sold

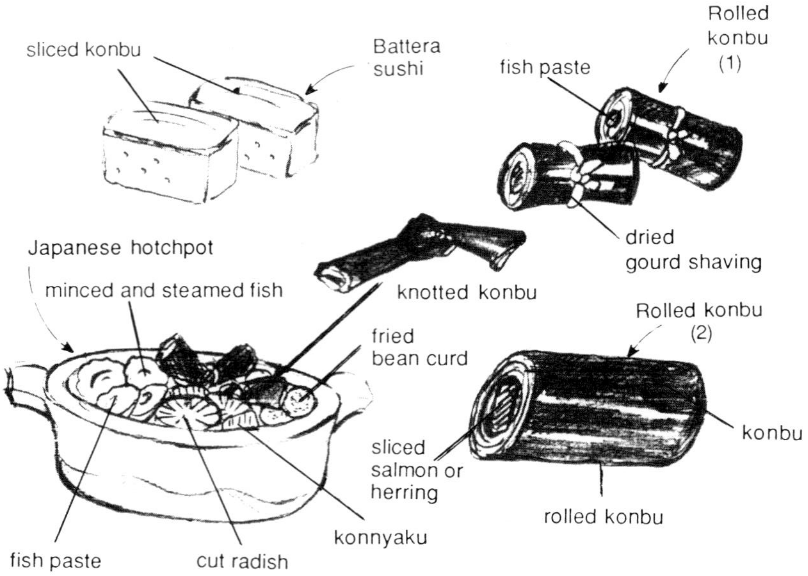

Figure 11 Using konbu

Figure 12 A screening machine for hoshi hiziki, which gives higher quality products.

through retail dealers or supermarkets as the seasoned product which was described above. Fairly large amounts of hiziki are imported from Korea in the form of hoshi hiziki. It is sold via Japanese import traders to processors who operate mostly in the Ise district of Mie Prefecture.

5. PRICE

In 1985 the price of hoshi hiziki was 1 200-1 300 yen for both domestic and imported products. The retail price was around 3 000 yen per kg, almost equal to that of konbu.

6. COOKING

Hoshi hiziki is usually cooked with soybean or fried bean curd and with some vegetables such as sliced carrot (Figure 13). This kind of hiziki is also available ready-made. Quite recently, hoshi hiziki has become available as a powder and, like konbu powder, is mixed into noodles and buckwheat noodles.

WAKAME

1. SOURCES

The wakame products in Japan are almost exclusively made from <u>Undaria pinnatifida</u>. This brown alga is distributed in the western and southern parts of Hokkaido, on almost the whole Pacific coast of Honshu (except the coasts of the Seven Islands of Izu and Koochi Prefecture), all coasts of Seto-Naikai (Inland Sea) and almost the entire coast bordering the Japan Sea and the Korean Peninsula.

For Japan, the yields of wakame during the past few years and the amounts of imported material are shown in Table 13 in the Statistical Year Book, 1983 (Statistics and Information Department, 1985). The yield of natural wakame has decreased during this time and it has been replaced by cultivated wakame. China is estimated to produce about 12 000 metric tons (wet weight) annually; the amount is very small compared to the Chinese production of konbu, but part of it is exported to Japan. In 1985 Korea produced nearly 300 000 metric tons (wet weight) of which about 80 000 metric tons were exported to Japan.

2. PREPARATION OF FOOD PRODUCTS

A. Suboshi wakame

The fronds of raw wakame are washed first with seawater and then with fresh water after harvesting, and they are cut into two similar halves by removing the midrib. They are dried in the sun or a hot-air dryer. The product is called suboshi wakame. It is sold in this form.

Table 12

Main composition of hoshi hiziki and the contents of other minor components on a dry weight basis

	Protein (%)	Lipid (%)	Carbohydrate (%) Nonfibrous	Fiber	Ash (%)
	12.3	1.5	54.4	10.7	21.2

Mineral (%)		Sterol (mg%)		Vitamin (mg%)	
Ca	1620	24-methylene cholesterol	1.8	Carotene	0.66
P	116	Fucosterol	48	Retinol potency (IU)	371
Na	1620			B_1	0.012
K	5093			B_2	0.17
Mg[a]	654			Niacin	2.16
Fe	63.7			C	0
Cu[a]	2.11				
Zn[a]	2.6				
Mn[a]	1.94				
Co[b]	0.014				
Ni[c]	2.8				

a/ Innan et al., 1985

b/ Ishii, Suzuki and Koyanagi, 1978

c/ Converted from the data for wet frond by Ishikawa, 1984;

and others: Standard Tables (Science and Technology Agency, 1982)

Table 13

Yields of wakame in Japan and its imported amounts during the past several years (x 1 000 tons, wet)

Year	Culture[a]	Wild[a]	Imports from Korea[b]	Imports from China[b]
1980	114	16	-	-
1981	91	14	26.962	0.016
1982	118	12	23.357	0.513
1983	113	10	24.032	0.951

a/ As raw product   b/ As boiled and salted product

B. Haiboshi wakame

The suboshi product is often faded in colour and has soft tissues because of autolysis in the frond. To prevent this, wakame fronds are mixed with wood ash or straw ash in a rotary mixer soon after harvest and dried in the sun for 2 or 3 days, followed by storage in a plastic bag in the dark. After appropriate periods (usually 1 to 6 months) the fronds are taken out and washed with water to remove adherent ash and salt. Then the midrib is removed and the rest of the fronds are dried in a drying room. They are packed in appropriate sizes and quantities for sale.

C. Blanched and salted wakame

Raw wakame fronds are heated soon after harvest at $80°C$ for about one minute and cooled quickly with water. The fronds, now a vivid green, are mixed with salt in a ratio of 3:10 (w/w) in a machine. They are preserved in a tank for 24 hours, then packed in a net bag to remove excess water. The product is called **yudoshi-enzo wakame** and is stored in a cold room at $-10°C$ for sale. This product is a major commercial form of wakame at present.

D. Cut wakame

The yudoshi-enzo wakame is desalted by washing with fresh water, centrifuged to remove excess water, then cut mechanically into small pieces and dried in a rotary type of flow-through dryer. These cut fronds are sorted into uniform sizes by sieving, any faded fronds (if present) are removed and then appropriate amounts are packed in a bag of plastic film, ready for sale (Figures 14 and 15).

3. NUTRITIVE VALUE

As with konbu and hiziki, wakame contains a large amount of nonfibrous carbohydrate which is composed mainly of fucoidan and water-soluble and -insoluble alginates. These acidic carbohydrates are very effective not only as a dietary fiber but they also have particular pharmacological effects on experimental animals, as described earlier for konbu. For example they lower the blood cholesterol level and show antibloodcoagulation, antilipemia and even antitumor activity. In addition, wakame contains a relatively large amount of fucosterol, on average more than that in konbu and hiziki. The pharmacological function of fucosterol has been described in the corresponding section on konbu.

The main constituents of various wakame products are shown in Table 14 and the contents of vitamins and sterols are in Table 15. The mineral contents of wakame products are shown in Table 16. Note that the quantities of microelements such as iron, magnesium and zinc are larger than those of konbu and hiziki. The contents of calcium

Figure 13 Using hiziki

Figure 14 Sieving and screening of cut wakame to obtain a product as homogeneous as possible and to remove any pieces of a lower quality, such as faded fronds

and phosphorus in haiboshi and cut wakame are relatively high while yudoshi-enzo wakame has a high sodium content. This high sodium content must be taken into account when it is cooked in food.

The contents of free and bound amino acids are shown in Table 17. The bound amino acids (that is the amino acids in protein) are noticeable in that they show relatively high amounts of neutral amino acids such as alanine, leucine, threonine and valine in addition to an almost equal level of dibasic amino acids (aspartic and glutamic acids). By contrast, in the case of the free amino acids, the amounts of alanine, glycine, proline and serine are far greater than the dibasic amino acids; this is the reverse of the free amino acid ratios in konbu and is a reflection of the different tastes of wakame and konbu.

It must be pointed out here that air dried (suboshi) wakame retains all vitamins in almost equal amounts to those of raw fronds and that they are relatively rich in B group. However a large proportion of these vitamins is lost in other processed products.

In contrast to konbu and hiziki, the major fatty acid constituents of the lipids of wakame are palmitic and oleic acids, in both thallus and sporophyll, although the total lipid content of the latter is as much as 2-3 times larger than that of the former (Table 14). Note that the sporophyll lipid contains 4.3% EPA.

4. MARKETING

Wakame which has been harvested by fishermen follows the marketing chains shown in Figure 16.

5. PRICE

In Japan during the last few years, the retail price of suboshi (sun dried) wakame was very unstable, as with suboshi konbu, varying between 1 800-3 000 yen per kg. The cost of the blanched and salted product (yudoshi-enzo wakame), the main commercial product, is 600-800 yen per kg. China exports wakame to Japan and in 1983 the export price was US$ 0.88 per kg. Korea also exports to Japan, in 1985 at a price of about 180 yen per kg of the blanched and salted product (yudoshi-enzo wakame).

6. COOKING

In Japan blanched and salted wakame is quickly desalted in water and used for various kinds of soups (Figure 17). Recently, different kinds of processed wakame goods have been manufactured and marketed as instant foods. The wakame products in Korea are mostly cooked for soup, as in Japan except that more wakame is used so that the soup has a far thicker appearance.

Table 14

Main composition of various products of wakame (g/100 g of sample)

Product	Water	Protein	Lipid	Carbohydrate Nonfibrous	Fiber	Ash
Raw wakame, whole	90.4	1.9	0.2	3.8	0.4	3.3
Merely dried (suboshi) wakame, whole	13.0	15.0	3.2	35.3	2.7	30.8
Blanched and salted (yudoshi-enzo) wakame	52.6	4.1	0.5	9.0	0.5	33.3
Stipe and midrib (kuki) wakame	90.5	0.8	0.1	3.2	0.7	4.7
Sporophyll, whole	12.4	11.6	2.9	41.6	3.7	27.8
Sporophyll without stipe	8.4	13.4	7.1[a]	30.0	2.9	38.6
Cut (katto) wakame	7.2	23.7	2.7	37.5	5.2	23.7
Ash-admixed and dried (haiboshi) wakame	13.8	18.1	1.7	41.3	3.6	21.5
Blanched and salted (yudoshi-enzo) wakame, desalted	91.8	0.6	0.1	6.8	0.1	0.6

[a] A recent analysis showed that the lipid contains 4.3% EPA (unpublished data)

Figure 15 The wrapping machine for cut wakame

Table 15

Vitamins and sterols of various products of wakame
(per 100 g of sample)

Product	Water (g)	Carotene	Retinol potency (IU)	B$_1$	B$_2$	Niacin	C	24-methlene cholesterol	Fucosterol
Raw wakame, whole	90.4	1400	780	0.07	0.18	0.9	15	19	62
Merely dried (suboshi) wakame without midrib	13.0	3300	1800	0.30	1.15	8.0	15		
Blanched and salted (yudoshi-enzo) wakame	52.6	840	470	0.03	0.07	0.2	0		
Stipe and midrib (kuki) wakame	90.5	210	120	0.02	0.05	0.8	14		
Sporophyll, whole	12.4	1900	1100	0.31	0.61	16.5	24		
Sporophyll without stipe	8.4	-	-	-	-	-	-	-	140[a]
Cut (katto) wakame	7.2	5630	3130	0.11	0.07	0.13	0		
Blanched and salted (yudoshi-enzo) wakame, desalted	91.8	49	27	0.03	0.01	0	0		
Ash-admixed and dried (haiboshi) wakame	13.8	1400	780	0.07	0.02	0.9	27		

a/ Recent unpublished data (1986)

Table 16

Minerals of various products of wakame (mg% of samples)

Product	Water[a] (%)	Ca	P	Fe	Na	K	Mg)[2b]	Co	Ni	Cu	Zn	Mn)[3c]
1. Raw wakame, whole	90.4	1042	375	7.3 (5.4)	6354	7604		(0.02)	(0.46)	(0.34)	-	(1.0)
2. Merely dried (suboshi) wakame without midrib	13.0	1103	460	8.0	7011	6322	1172	-	-	0.25	3.1	0.8
3. Blanched and salted (yudoshienzo) wakame	52.6	401	200	5.9	27426	527						
4. Blanched and salted (yudoshienzo) wakame, desalted	91.8	244	122	2.5	2805	61						
5. Stipe and midrib (kuki) wakame	90.5	737	242	2.1	5474	20000						
6. Sporophyll, whole	12.4	753	559	9.5	3653	10046						
7. Sporophyll without stipe	8.4	677	361	28.0	3663	8658	616	-	-	0.4	2.1	2.3
8. Ash-admixed and dried (haiboshi) wakame	13.8	1856	917	24.2	6497	360						
9. Cut (katto) wakame	7.2	1099	400	19.2	8157	337						

Product 1: figures in round brackets, (), from: Ishii, Suzuki and Koyanagi, 1978
Product 2: Mg, from: Analysis Certificate (Japan Food Research Laboratories, 1985)
3[c] : from Innan et al., 1985
Products 1-6: 1[a] and 2[b] from: Standard Tables (Science and Technology, 1982)
Product 7: from: Tanaka, Mori and Ishiwaza, 1984
Product 8: from: Japan Frozen Foods Inspection Corporation, 1984
Product 9: from: Standard Tables (Science and Technology Agency, 1982), and Analysis Certificate (Japan Food Research Laboratories, 1985)

Table 17

Free and bound amino acids composition of wakame on a dry weight basis

Amino acid	Free (mg%)	Bound (g% of protein as Total N x 6.25)
Alanine	612	4.48
Arginine	36.5	3.04
Aspartic acid	5.4	5.92
Glutamic acid	89.8	6.56
Cystine	3.4	0.93
Glycine	455	3.68
Histidine	2.1	0.50
Isoleucine	11.2	2.88
Leucine	19.6	8.48
Lysine	34.6	3.68
Methionine	1.7	2.08
Phenylalanine	9.2	3.68
Proline	156	3.04
Serine	131.4	2.56
Threonine	90.3	5.44
Tryptophan	5.8	1.17
Tyrosine	10.1	1.60
Valine	11.1	6.88

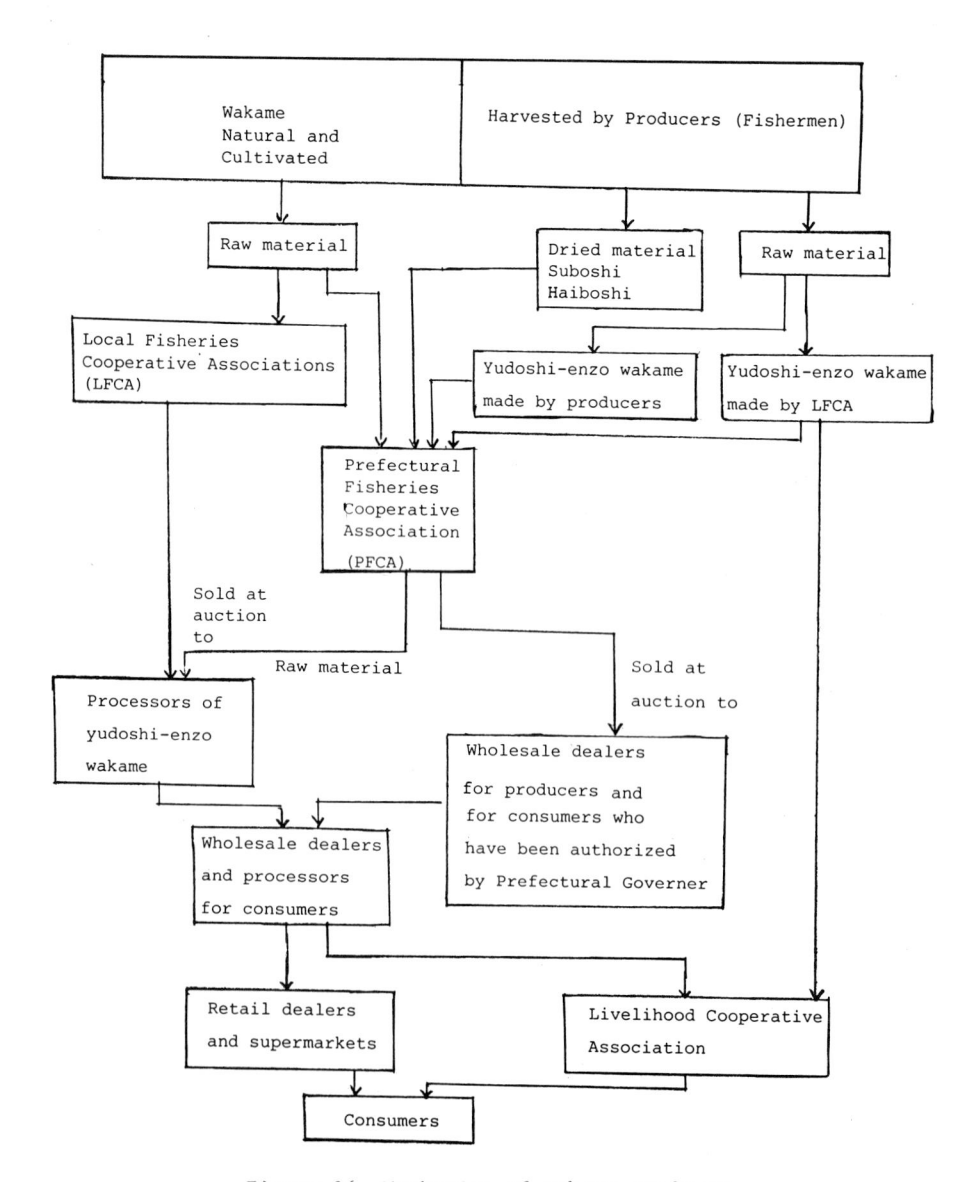

Figure 16 Marketing of wakame products

Figure 17 Using wakame

REFERENCES

Abe, S. and T. Kaneda, 1975. Studies on the effect of marine products on cholesterol metabolism in rats. 11. Isolation of a new betaine, ulvaline, from a green laver Monostroma nitidum and its depressing effect on plasma cholesterol levels. Bull.Jap.Soc.Sci.Fish., 41:567-71

Dyeberg, J. and H.O. Bang, 1979. Haemostatic function and platelet polyunsaturated fatty acids in Eskimos. Lancet, 1979:433-5

Dyeberg, J., et al., 1978. Eicosapentaenoic acid and prevention of thrombosis and atherosclerosis? Lancet, 1978:117-9

Fujiwara, M., et al., 1984. Purification and chemical and physical characterization of an antitumour polysaccharide from seaweed, Sargassum fulvellum. Carbohydr.Res., 125:97-106

Furusawa, E. and S. Furusawa, 1985. Anticancer activity of a natural product, Viva Natural extracted from Undaria pinnatifida, on intraperitoneally implanted Lewis lung carcinoma. Oncology, 42:364-9

Hiramatsu, M., Y. Niitani and A. Mori, 1981. Effect of taurocyamine on taurine and other amino acids in animal body. Sulfur Amino Acids, 4:227-32

Horiguchi, Y., H. Noda and M. Naka, 1971. Biochemical studies on marine algae. 6. Concentration of selenium in marine algae and its importance as a trace metal for the growth of lavers. Bull.Jap.Soc.Sci.Fish., 37:996-1001 (in Japanese)

Hosogai, Y., Y. Naoi and T. Okada, (eds), 1978. Manual of toxic elements. Tokyo, Chuoohoki Shuppan Inc., p. 205 and p. 425 (in Japanese)

Innan, S., et al., 1985. Food contents of dietary fibers, minerals, cholesterol and fatty acids. Tokyo, Ishiyaku Shuppan Inc., pp.130-1 (in Japanese)

Ishii, T., H. Suzuki and T. Koyanagi, 1978. Determination of trace elements in marine organims. 1. Factors for variation of concentration of trace element. Bull.Jap.Soc.Sci.Fish., 44:155-62 (in Japanese)

Ishikawa, M., 1984. Unpublished observation at the Division of Marine Radioecology, National Institute of Radiological Science, Isozaki, Nakaminato, Japan

Japan Food Research Laboratories, 1985. Analysis certificate, No. 28090669 003, Japan Food Research Laboratories, Osaka Branch, Osaka (in Japanese)

Japan Frozen Foods Inspection Corporation, 1984. Kobe Branch, Kobe. Examination certificate, K 59, 101, No. 117 (in Japanese)

Kagawa, A., (ed.), 1983. Amino acid composition in Japanese foods. Standard tables of food composition in Japan. Tokyo, Japan, Resources Council, Science and Technology Agency, pp. 253-9 (in Japanese), 4th. ed.

Kanazawa, A., 1963. Vitamins in algae. Bull.Jap.Soc.Sci.Fish., 29:713-31

Korea, Ministry of Agriculture and Fisheries, 1983. Yearbook of fishery statistics, 1983. Seoul, Republic of Korea, Ministry of Agriculture and Fisheries

_____, 1985. Yearbook of fishery statistics, 1985. Seoul, Republic of Korea, Ministry of Agriculture and Fisheries

Loose, G., H.A. Hoppe and O.J. Schmid, 1966. Meeresalgen für die menschliche Ernaerung. Bot.Mar., 9.Suppl.46 p.

Matano, K., 1959. Studies on specific components in foods: studies on precursor of dimethyl sulfide of Ulva sp. and Monostroma sp.; antigastric ulcer substances in algae. Annu.Rev. Natl.Inst.Health, Tokyo, 12:241-2 (in Japanese)

Miyashita, A., 1974. The seaweeds: the cultural history of material and human being. Tokyo, University of Hosei Press (in Japanese)

Mori, H., et al., 1982. Sugar constituents of some sulfated polysaccharides from the sporophylls of wakame (Undaria pinnatifida) and their biological activities. In Marine algae in pharmaceutical science, edited by H.A. Hoppe and T. Levring. Berlin, Walter de Gruyter, Vol.2.110-21

Nakamura, S., et al., 1968. Isolation and identification of nucleotides from several marine algae. Bot.Mag.Tokyo, 81:556-65 (in Japanese)

National Federation of Nori and Shell Fishes, 1985. Cooperative Associations. Nori Times, (1838):p. 1 (in Japanese)

Noda, H., 1971. Biochemical studies on marine algae. 3. Relation between quality and inorganic constituents of hoshi nori (dried Porphyra yezoensis). Bull.Jap.Soc.Sci.Fish., 37:35-9 (in Japanese)

Noda, H., Y. Horiguchi and S. Araki, 1975. Studies on the flavor substances of nori, the dried laver Porphyra sp. 2. Free amino acids and 5'-nucleotides. Bull.Jap.Soc.Sci.Fish., 41:1299-1303

Noda, H., et al., 1981. Sugars, organic acids and minerals of nori, the dried laver Porphyra sp. Bull.Jap.Soc.Sci.Fish., 47:57-62

Okubo, T., 1985. The trend of production and consumption of edible seaweeds in Japan. Up to Date Food Process.,Tokyo, 20:30-2 (in Japanese)

Ozawa, H., Y. Gomi and I. Otsuki, 1967. Pharmaceutical studies on laminin monocitrate. Yakugaku Zasshi (J.Pharm.Soc.Jap.), 87:935-9 (in Japanese)

Sakagami, Y., 1983. Isolation of phorphyosin and verucoysin from edible seaweeds. In Biochemistry of marine algae and their application, edited by the Japanese Society of Scientific Fisheries. Tokyo, Koseisha Koseikaku Inc., pp.90-100 (in Japanese)

Sekiguchi, M. and M. Maeda, 1983. Master thesis, Department of Biochemistry, Saitama University (in Japanese)

Shimonaka, M., et al., 1984. Successive study on the production of plasminogen activator in cultured endothelial cells by phytosterol. Thrombosis Res., 36:217-22

Science and Technology Agency, Resources Council, 1982. Standard tables of food composition in Japan. Tokyo, Ishiyaku Shuppan Inc. (in Japanese) 4th. ed.

Statistics and Informational Department, (ed.), 1985. Statistical year-book of fisheries and marine aquaculture, 1983 (60th). Tokyo, Ministry of Agriculture, Forestry and Fisheries, 1985 (in Japanese)

Takahashi, T., 1933. Studies on the chemical composition of brown algae, 28:36-44 (in Japanese)

Takemoto, T., K. Daigo and N. Takagi, 1965. Studies on the hypotensive constituents of marine algae. 3. Determination of laminine in the Laminariaceae. Yakugau Zasshi (J.Pharm.Soc.Jap), 85:37-40 (in Japanese)

Tanaka, M., H. Mori and K. Ishizawa, 1984. Food-chemical studies on seaweeds. 2. Composition of blade and sporophyll of Undaria pinnatifida (wakame). J.Tokyo Kaseigakuin Coll., 24:45-47 (in Japanese)

Tseng, C.K., 1984. Modern seaweeds of China. Beijing, Science Press, 316 p.

Tsuji, K., et al., 1981. Effect of dietary taurine on bile acid metabolism in hypercholesterolemic rats. Sulfur Amino Acids, 4:111-9 (in Japanese)

_____, 1983. Hypocholesterolemic effect of taurocyamine or taurine on the cholesterol metabolism in white rats. Sulfur Amino Acids, 6:239-48 (in Japanese)

Yamada, M., 1985. Determination of folate contents in dry seaweeds by microbiological assay method. Vitamins, 59:509-15 (in Japanese)

Yamamoto, I. and H. Maruyama, 1985. Effect of dietary seaweed preparations on 1,2-dimethylhydrazine induced intestinal carcinogenesis in rats. Cancer Lett., 26:241-51

Yamamoto, I., et al., 1982. Antitumour activity of crude extracts from edible marine algae against L-1210 Leukemia. Bot.mar., 25:455-7

_____, 1984. Antitumour activity of edible brown seaweeds against L-1210 Leukemia. Hydrobiologia, 116/117:145-8

Anon., 1984. Annual report of nori. Appendix. Provision News, Tokyo, 1984:Appendix, pp.144-55 (in Japanese)

Tipo-lito-SAGRAF - Napoli